FORKS IN THE ROAD
A LIFE IN PHYSICS

FORKS IN THE ROAD
A LIFE IN PHYSICS

STANLEY DESER
BRANDEIS UNIVERSITY, USA & CALTECH, USA

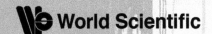
World Scientific

NEW JERSEY · LONDON · SINGAPORE · BEIJING · SHANGHAI · HONG KONG · TAIPEI · CHENNAI · TOKYO

Published by

World Scientific Publishing Co. Pte. Ltd.

5 Toh Tuck Link, Singapore 596224

USA office: 27 Warren Street, Suite 401-402, Hackensack, NJ 07601

UK office: 57 Shelton Street, Covent Garden, London WC2H 9HE

Library of Congress Cataloging-in-Publication Data

Names: Deser, S. (Stanley), author.

Title: Forks in the road : a life in physics / Stanley Deser.

Description: New Jersey : World Scientific, [2022] | Includes index.

Identifiers: LCCN 2021028052 (print) | LCCN 2021028053 (ebook) |
    ISBN 9789811234187 (hardcover) | ISBN 9789811235665 (paperback) |
    ISBN 9789811234194 (ebook for institutions) | ISBN 9789811234200 (ebook for individuals)

Subjects: LCSH: Deser, S. (Stanley) | Physicists--United States--Biography.

Classification: LCC QC16.D48 A3 2022 (print) | LCC QC16.D48 (ebook) |
    DDC 530.092 [B]--dc23

LC record available at https://lccn.loc.gov/2021028052

LC ebook record available at https://lccn.loc.gov/2021028053

British Library Cataloguing-in-Publication Data

A catalogue record for this book is available from the British Library.

Background Image: "Between what I say and what I keep silent" by Elsbeth Deser.
    The painting is in the collection of the Harvard Fogg Art Museum.
    Credit Line: Harvard Art Museums/Fogg Museum, The Jorie Marshall Waterman '96 and
    Gwendolyn Dunaway Waterman '92 Fund
    Copyright: © Abigail Deser
    Accession Year: 2007

Cover art design by Abigail Deser & Emilia Peters.

For any available supplementary material, please visit
https://www.worldscientific.com/worldscibooks/10.1142/12205#t=suppl

Editor: Abigail Deser

Typeset by Stallion Press
Email: enquiries@stallionpress.com

When you come to a fork in the road, take it.

      Yogi Berra (and the basic law of Quantum Mechanics)

Dedicated to my wife, children, grandchildren, and family members, deceased and living. And to all victims of the Holocaust. No man is an island.

CONTENTS

INTRODUCTION

If writing an autobiography requires examination of one's motives, writing a scientific one requires examination of one's sanity. It is an almost impossible task, because one can never satisfy the lay reader and the scientist simultaneously. I believe that my motives are not entirely due to vanity or a desire for immortality, and that I am not (yet) certifiable. As far as I can vouch for them, my intentions are to provide a map of the many ways one becomes a theoretical physicist and what travails are involved, so as to either encourage or discourage future candidates; also to provide, to some extent, a mirror to those of us already engaged, who will perhaps recognize much that I mention. I will not intrude much technicality into this account, but some appendices provide that flavor. I was lucky enough to live in a time that enabled me to meet many of the legends of our field, under various circumstances, and I still remain aware of what is going on at present. I will try to draw a picture of what our profession's life and times were and are like. As I write these lines, I hear Julian Schwinger's (my late thesis professor) wry dictum upon taking up the piano at a late age: "if it's worth doing, it's worth doing badly."

It is a truism that accidental circumstances often shape one's fate. I will cite two: World War I lasted far beyond 1914–1918, all the way to 1922 where I was born. The Austrian and Russian empires were followed by the Poles, Ukrainians, Czechs, White and Red Russians, who unleashed alternating waves of violence in the region. When my mother was twelve, as the daughter of rich merchants, she was kidnapped at gunpoint during a Bolshevik wave. [See Isaac Babel's 1920 Diary for a detailed description of this violent era.] Life was cheap then and she might well have been murdered had not the next army liberated her. Life-altering events can also happen under the most idyllic circumstances. In the mid-1950s, my wife and I were driving on an empty rural Swedish road on a summer's

1

day, when the steering wheel stopped responding, as if it had hit black ice. After a couple of 360 degree spins, we ended up in the ditch—fortunately right side up—as an enormous gasoline truck thundered past. After a very slow ride to the nearest garage, we were shown the reason: the shaft from the steering wheel to the axle was secured by a single giant wing nut that had reached its final turn. It was hardly the only flaw in our Morgan: Its electrical system was by Lucas, known as the Prince of Darkness to British motorists, as we could amply corroborate.

The aleatory nature of life is, of course, a basic feature of almost all literature. At the end of Proust's *Remembrance*, a demi-mondaine's daughter turns into a duchess and a vulgar salonniere into a princess. Tolstoy's *War and Peace* describes Napoleon's 1812 degradation and Kafka's is perhaps the best-known example of metamorphosis. While my account does not deal with such dramatic reversals, it provides its own illustration of life's vagaries. This is not to say that one cannot influence one's own destiny; as the bromide goes, there is a confluence between luck and preparedness.

EARLY DAYS: ROOTLESS COSMOPOLITAN

I was born, and grew up in, a particularly tumultuous and tragic time. Unlike most of the people of my generation in Europe, I was extremely lucky. My parents were one step ahead of annihilation at any given moment. Let me strongly emphasize that I was never directly exposed to that era's unspeakable horrors.

My arrival was on March 19th, 1931, some five years after QM and fifteen after GR, in Rovno (now the Ukrainian Rivne), a Polish city near the Soviet border. My parents were reasonably well-off and very health-conscious about me, as was standard for their class in the early 1930s. Thus, I was told that I was put outside in a baby carriage, once near a record 40 below zero (take your choice—Fahrenheit or Centigrade) in order to inure me to nature, and was apparently fed progressively, although I did catch a mild case of tuberculosis from a wet-nurse. The only event that I remember, or imagine that I do, is of my grandfather dropping me down a flight of stairs on my head when I was about a year and a half. I will not speculate on whether that had a positive or negative effect on my intellectual abilities. I was a sickly child; tonsillectomy was the panacea, so my parents took me to a great specialist in Lwow. Without warning, he brutally chloroformed me. The post-op consisted of lots of ice cream, of which the Polish was the best, a small consolation.

I lived a happy little life in a relatively well-off environment until the summer of 1935. My parents, who had each graduated from French universities and hated Poland's anti-Semitic and repressive regime, decided to emigrate to what was then Palestine. My father had an advanced degree in chemical engineering from Toulouse and was employed by Alfa-Laval, a Swedish agricultural instrument company, inspecting their products all over Poland. With the start of the Depression, the company shrank its foreign operations and he was made redundant. It still seems strange to me

3

that they chose to leave Poland, as his father's earlier death had left his mother alone in a small town near Rovno, while my mother was her family's favorite. But I suspect emigration was the only choice, given Poland's economic state. However, Palestine was an underdeveloped country and 1935 was a bad time to arrive, during one of the endemic, but particularly violent, episodes of the Arab–Zionist settlers' civil wars under the British Mandate.

My central memory of the long train ride from Rovno to our ship was the first shock of the sea. The train emerged from a long tunnel: suddenly, there was the blue Adriatic. I realized much later that this is the spot where the train passes the ICTP (International Centre for Theoretical Physics) that Abdus Salam founded in Trieste, quite a few decades later. Trieste had a large aquarium, which did not, however, awaken any scientific interest in me. The boat trip to Haifa was of course exciting to a small child suddenly presented with a whole new environment, far from the rigors of the Continent. I remember a grueling summer in Tel Aviv, which was then almost a Third World village, where I acquired a variety of tropical diseases that covered me with sores, probably from the open street sewers in our quarter (Nevei Shaanan). The standard remedy was immersion, I presume including the heel, in the Dead Sea. To my, and my grateful parents', amazement, it worked. Thanks to the minerals, I walked out entirely rid of my blemishes.

After only a few months, it became clear to my parents that the prospects for establishing themselves in Palestine were non-existent. Already in 1934, my father had set out for France on his own as the French Prime Minister Léon Blum had a liberal view of immigrants, but he did not feel confident enough to bring us there then; now he had no choice. Our boat from Haifa went to Marseille via Alexandria, whence we took the standard side trip to the Pyramids, Sphinx and all, still vivid to me now. We lived in Paris from the late summer 1935 to June 12th 1940, an ominous date. My formal education began, and almost ended, in the French public and private school systems. The kindergarten, known as the *École Maternelle*, was the polar opposite of its name. We were crowded, about 43–44 kids per class, all wearing black aprons, presided over by a formidable presence whose main pleasure was using her ferrule, but not for measuring. She loved to quell any disturbances and menaces in her class of dangerous five year-olds with it. Another wrinkle in the system was the weekly reseating of the students according to our perceived performance in the previous one. I always ended up around 43rd or 44th in line; I may have reached 42nd once. This technique of course achieved the wrong end, because all we stupid children were furthest away from the font of wisdom. We were taught various bits of French patriotic lore, some strange attempts at science, like using a scale and not much else. Needless to say, I didn't benefit from the storied French educational system. Luckily, I had become used to learning languages: Polish, then Hebrew, so now French was added. I also learned to read; it suddenly came, and the first thing I devoured was *Gulliver's Travels* (I suppose in a children's edition) while we sat in cafes on Sundays. This was a treat made possible by my father's small chemical enterprise that he had managed to establish successfully this time. Another special memory is of the 1937 Paris World's Fair, that my maternal grandparents were able to visit with us, availing themselves of the packaged tours that were sold all over Europe. I proudly acted as the big city guide to my country visitors.

All was not peace and serenity in those pre-war years, however: my mother took me to spend the summer of 1938 with her family in Rovno. This required long, sooty train journeys across Germany and Poland, but the Polish countryside was so unspoiled and healthy that it seemed worth the voyage. Indeed, it was a glorious summer in the rural outskirts of Rovno—that is, until it was time to go back to Paris. The Munich crisis had just arisen, so crossing the Reich was forbidden, with war imminent. The crisis was very temporarily averted by the Chamberlain–Ribbentrop pact, the first step in ceding the Czechs to Hitler. We left immediately, ironically saved by this first of many dictator-led salvations. Germany was in full wartime production, our train compartments were filled with Nazi uniforms. My mother was very panicky until we reached the French border; only then did I begin to understand the danger we were facing.

Of the last peacetime year I have little recollection, except my parents' willingness to let the seven-year-old me roam the Paris streets alone, at least on Saturday afternoons, when I would walk quite far to see children's movies near the Opera. Our material life also improved enough that my parents were able to accumulate some savings that they instantly converted into gold coins and kept in a bank safe, the standard emigre's protection. In September 1939, World War II broke out. All public schools were closed for fear of German air raids. That was the "phony war" period when seemingly nothing happened.

My parents decided to put me, not that they had much choice, into a horrible little private school; there were two of us in my class. It wasn't really a school, just a woman and a few stray kids. At least I wasn't seated at number 44. I still remember when she decided to hand out medals, as they did in the public schools, to be returned at week's end. First place looked like a Napoleonic Marshal's award, second was rather more modest and third was a shapeless piece of tin. The first time she handed them out I came home proudly bearing the third place medal and my parents asked, "But aren't there only two kids in your class?" Soon, this school also closed. Only the Catholic ones remained; France was a secular country so these schools had no special status and could choose to stay open. So my next one was Jesuit, but the Principal assured my parents no one would learn my heretic background; of course, it was immediately apparent as I didn't have to participate in prayers. The children were clearly (pre-)Vichy-minded, and treated me accordingly, so I learned about bullying at an early age. We were not taught much in that school either. This was a terribly stressful period for everyone, my parents included. What saved me were the beautiful original editions of Jules Verne I borrowed from the public library. I read them surreptitiously under my desktop until I was caught in early June. The Principal had me in and said that if I wanted to get those books back to the library, my parents would have to come in and discuss it with her. So one of the benefits of the German invasion was evading that interview. *Habent sua fata libelli!*

We lived on Rue Lafayette, a broad boulevard on the north side of Paris and had begun to see the exodus of Dutch and Belgian refugees earlier, in mid-May. Some were on foot, some in cars and other conveyances, heading south. Somewhere between the 10th and the 12th of June, it became clear that the Germans had started into France. We owned a tiny car, a Simca 5, that looked like Mickey Mouse. Neither of my parents could drive; my uncle, who lived with us, was the only one. They finally realised the danger and decided to leave everything; I rushed with my father to empty our safe. That evening, my mother sewed the coins into a belt of towels, a much-practiced maneuver of refugees, while the rest of us packed a few belongings. Even I understood the gravity of the situation. After a restless night, the four of us jumped into the car with our small bags. The Nazis rolled in that afternoon. We were part of the enormous exodus from Paris, swelled by the Dutch and Belgians, that began just as the Germans hit Paris from the north. Below Paris, we veered off onto smaller byroads, occasionally being strafed, which is not as much fun as it sounds. My father, being a chemist, was able to improvise fuel at pharmacies since of course many gas stations had gone dry. We skirted around cities where the bombings and traffic jams were especially intense. We had a terrible incident during one of the bombings, one that might have done us all in: all cars came to a halt. My mother and I ran out of the car to find shelter, as did many others. We did not realize that the car queue was suddenly moving again, and that my uncle and father had to start driving. We were all completely out of our minds as to how we would be reunited in this chaos. Luckily, this was a one block, one street village, so we eventually found each other as they were able to pull out of the line.

About a week from Paris, we reached Bayonne and Biarritz, very close to the southwest Atlantic Spanish border, having by-passed Orleans and Bordeaux. [Nowadays, the TGV takes a couple of hours to get to Bordeaux and autoroutes are omnipresent.] We slept and ate where and how we could. At that point we benefited from a double miracle, one finally recognized by a wider public. Aristides de Sousa Mendes, the Portuguese consul in Bordeaux who oversaw the French southwest region, foreseeing the coming slaughter, unilaterally decided to issue transit visas to all who were waiting in his sub-consulates. We were part of those crowds, worriedly putting our (life and death) passports into a basket that was hauled up through a window and thankfully, back down with a life-saving transit visa. Consul Sousa Mendes' heroic actions were forbidden by the Salazar dictatorship in Lisbon and he suffered bitterly for them the rest of his life. Only in recent years has he been recognized for the more than thirty thousand lives he saved. Indeed, the Portuguese government has finally given him full recognition as I write these lines. On the basis of his visa, one could get a Spanish transit one, just before Franco followed suit and shut his border. Sousa Mendes' counterpart, the equally heroic Spanish Ambassador de Callejon, was similarly punished—by Franco—for issuing

30,000 unauthorized visas. Both are rightly remembered in Yad Vashem. My father, being a Polish citizen, also had to get an exit permit, since he was still of age for the Polish army! This had to be secured from the French military authorities since they were allies. Each of these proceedings was totally chaotic and anxiety-filled, as may be imagined.

Armed with all the necessary documents, we decided to get out as fast as possible, at all costs. We simply parked our car at the border town of Hendaye's railway station, where it must have accumulated many tickets. I still remember its "mineralogical" license plates, (RM8910) as the French called them: licensing in France was initially delegated to the Bureau of Mines for obscure Cartesian reasons. Hendaye was separated from Spain by a little river, the Bidassoa. My parents had wisely brought some kilos of sugar, which they made quite visible, because export of food during the war was prohibited. As we passed through French customs, we duly declared them and the officials duly confiscated them and because of this smart decoy, didn't look any further. Unfortunately, though on a smaller scale, the gold coins which had been draped around my waist had been secured by safety pins, one of which had opened as we were passing through customs. As in the old fable of the Spartan boy and the fox, which I had read as a child, I was forced to be brave about this pain but the memory is indelible.

The French surrender, or Armistice, was around June 21st, a bit over a year after the Spanish Civil War ended. Spain was still a totally devastated country, with no food. In Irun, across the French border, we boarded a dilapidated train that went southwestwards through Spain to the nearest Portuguese border town, Vilar Formoso. This two-day trip was not very pleasant either, but we got to peaceful Portugal. We later discovered that ours was the last train out of France before the Armistice; once more saved in the nick of time. Indeed, as we crossed that border river on the train, I saw a German soldier standing at one of the watchtowers: the Wehrmacht had already occupied the coast of France. Our Spanish journey was entirely fueled by a single, almost inedible bread we bought through the window as the train stopped early one morning: a local peasant appeared, and we had to part with a British gold sovereign for it, a rather high price. When we reached the Portuguese border, families were sent, seemingly at random, to towns (not camps) all over the country. Ours was called Figueira da Foz (very few people drew Lisbon, where all the consulates and embassies were) which pretended to be the Monte Carlo of Portugal. It certainly was not *the* Monte Carlo, nor even Estoril; indeed, it was probably the least picturesque place in the country. We rented a little apartment in the miserable available housing. The poor inhabitants were extremely welcoming to us refugees. Perhaps the fact that many of them had Jewish roots from five hundred years earlier predisposed them towards us.

Once settled into our new quarters, the future's uncertainties began to loom: Portugal was a potential Nazi ally, and in any case might decide to deport us liberal foreigners holding only transit visas. As the European war news was steadily getting worse, the possibilities of a long-term haven shrank. Indeed, we refugees ended up in many countries of Latin America, especially Cuba and Mexico, but even in such unlikely ones as Ecuador; some even stayed on in Portugal through the war, absent the ability to procure visas. Our own search for a visa involved endless nervous discussions amongst the adults and frequent visits to Lisbon's consulates, only to be relieved by visits to the drab casino. We children were shielded from all this to some extent. I was taught to swim and bicycle— badly: swimming lessons consisted of being lowered into a small river from a bridge, while wearing a leather belt held by the instructor who stood above (and may not have known how to swim).

To go to Lisbon required an easily-obtained permit that allowed one to stay there for a night or two; rarely did any of us kids go. Most of us had been educated, especially the other ones, in good Belgian and French schools. In general, the people who escape are the ones who are smart, well-off and quick on the draw. When school began, all parents were— rightly—so preoccupied that they didn't pay much attention to that either. We were an anarchic republic of all ages. The Portuguese schools' main aim was to keep the country illiterate, per Salazar's plan. Nothing beyond the 16th Century was taught lest it incite people to dispute the dictatorship. I still remember the main feature of every room: aside from a crucifix, there was a huge map showing the main Portuguese colonies, Angola and Mozambique. A map of Portugal, which is of course a tiny rectangle, completed the picture. There was also a map of the world as it was drawn by the great Portuguese navigators, who really were great, in the 15th and 16th Centuries. Then there would be some poem by the national bard,

Camoes, also from those ages. We were far from the 20th Century. While there was little meaningful pedagogy, we quickly learned the language by osmosis, as kids do.

I obviously can't complain about those eleven months in Portugal while waiting for a US visa. The State Department was extremely anti-Semitic. Cordell Hull, the Secretary of State did not want to admit "those people", even though there was a quota per country, and ours (Polish) was underused. Various excuses were found not to satisfy our request. In particular, one needed affidavits from US citizens declaring their ability to support us so we would not be a burden to the US social net (which was virtually non-existent at that time, apart from Social Security). My mother had an uncle in Brooklyn, who was relatively well-off. Jack Rosenberg was her mother's younger brother, who had come to the US in the early 1920s. He provided quite adequate affidavits, but even that was not enough. Things got stretched out; we had to show that we had enough money on our own not to be a burden to our US relatives. In the end, we were required to show up at the US consulate in Lisbon, specifically exhibiting what fortune we had, what means of support. This was not particular to our experience. Everyone in Figueira pooled their gold coins to make their holdings look impressive. Someone would then ride shotgun with the applicant to make sure he didn't escape (to where?) with all the gold after the interview. After this ruse, we were still required to get tuberculosis X-rays to show that we were not transporting this terrible disease.

My uncle was not so fortunate. The consulate decided that our affidavits could only support three people. He had to find another route. He left before us and I remember waving goodbye to him at the boat in Lisbon. His only choice was a transit visa to Shanghai, issued by the Chinese consulate in Lisbon. Of course, Shanghai was then under Japanese occupation and he had no intention of going there anyway. It did get him to Portuguese Mozambique, where he got off the ship and walked across the South African border. This could be done if you were able-bodied and willing to enlist in the South African army, which he did. He fought alongside the British army in Libya and Egypt and ended up in Palestine where he stayed the rest of his life.

We, on the other hand, had to procure passage to the US. That was not so easy; Pan American had established the Yankee Clipper, as we all know from movies of that era, a seaplane between New York, the Azores and Lisbon. We could not buy Clipper tickets, being mere immigrants: its seats were all reserved for important people. Instead, we had to pay even more to get steerage on one of the Portuguese Atlantic ships, three converted freighters that everyone who reached the US well remembers. The Portuguese quickly realized that we were a more valuable commodity than freight, so they installed triple cots in the hold, the only class. We left on the "Guinea" on May 10th 1941. To put that date in perspective, Germany's invasion of Russia was June 22nd, Pearl Harbor December 7th.

While still in Portugal, my parents had written to their families in Poland which, by then had been occupied by the Russians, who had annexed half of the country through the infamous 1939 Molotov-Ribbentrop Treaty. They wrote on Red Cross postcards that were used for communicating across warring countries, that we were safe in Portugal. We know those cards were received, because they replied, albeit briefly. Those were all the exchanges my parents had with their families before they were all slaughtered in the fall of 1941 in Rovno by Polish, Ukrainian and helpful German extermination forces. We only learned the details from survivors shortly after war's end when we had returned to New York, causing my parents to break down for quite a time.

AMERICA

There was a danger of U-boats, but Portugal was, if not an ally, at least formally neutral. Still, it was always conceivable that an eager U-boat didn't see or care what flag our ship was flying. So we zigzagged, and twelve days later dropped anchor, not at a Manhattan dock, but at a cheaper one on Staten Island. We had the classic view, as we sailed in at dawn past the Barnegat light to the promised land: the Statue of Liberty and Lower Manhattan. Before we were let off the boat, we had the sanitary inspection, each of us holding our enormous sheet of lung X-rays, which was carefully scanned. In the old Ellis Island days, the immigration inspectors arbitrarily renamed the newly landed. Our experience was not so drastic; we were allowed to make our own choice of first name. I was born Salomon, then underwent a change to Lucien in France because my parents felt it would ease my way. We were told Lucien would never do in the US, so among the options we were offered, Stanley seemed the least objectionable. My father's first name Nachman became Norman while my lucky mother kept Miriam.

A fleet of taxis, organized by HIAS, the Jewish refugee agency, loaded us onto the Staten Island ferry to Manhattan, where the affidavit providers could find their relatives. Uncle Jack, whom my mother hadn't seen in twenty years, took us to his house in Brooklyn. He and his wife Rae had two boys, one my age, the other a little older. By then, French was my native language, after Polish, Hebrew and Portuguese. I knew not a word of English except perhaps "chewing gum" (which was also a single French word "chewngum"), so we got along with sign language. The next day, my cousin hauled me off to Public School 186 as a trophy and I felt examined like an exotic animal. His class was studying "our friendly neighbors"; this was Brazil's turn, and they devoted their time to cutting out and displaying pictures of the country. My impression at the time was a rather negative one, since I couldn't see what, if anything, was being taught or learned. This impression continued in the remaining short time before school ended for the summer. My parents spoke Yiddish with the relatives, a language I had begun to learn as well, in Portugal. There was at last a feeling of being in safety that the three of us shared. Though I was included, I didn't really have much interest in the activities of my cousins and was more preoccupied with what fate would next dish up for me. It was clear to my parents and me that we were still in transit.

Within a few weeks of our arrival in New York, we were preparing to leave it. Most likely, Uncle Jack felt a bit uneasy at the prospect of our becoming long-term dependents and his solution was to send us to St. Louis (by Greyhound). My father was a chemist and St. Louis was the center of the US chemical industry, so presumably he could find work more easily there. In retrospect, I can think of two other factors: the Rosenbergs did not grasp that our flight was not due to economic pressures but was a desperate one for survival; less consciously, they may have felt that we were having it too easy compared to their own initial travails, and should also have to earn our way.

Like all the manufacturing cities in the US, St. Louis was invaded by workers coming from agriculture, particularly from the South, to the defense plants. The US was still at the tail end of the Depression; it was the war work that began to change that tide. Some distant, very aged and impoverished, relatives of my father were supposed to take care of us. Since they couldn't speak English themselves, they could hardly do so. We settled into a depressing apartment at the edge of the white-black divide, a deep one, which was shocking to us, coming from Europe. Little old black ladies would step off the sidewalk as I passed. Once I made the mistake of getting up to give a black woman my seat on the bus, which nearly caused a race riot. I was shunted off to the nearest junior/high School: Soldan High, which was genteel in a pre-war Tennessee Williams way (indeed, I recently discovered that he had attended it!). As an exotic animal, I was again well-treated and for some reason, bumped up to seventh grade, even if I didn't know English (it was still not much of a challenge). We spent a couple of years in St. Louis. My father found interesting work as a chemist rather easily, working under a White Russian emigre. I quickly became fluent in English and my parents less so, but they

13

managed. We were socially rather isolated through those years, nor did I make any school friends. I recall a summer spent at a YMCA camp in the Ozarks, and the vast quantities of watermelon consumed in local "melon gardens".

My father did not take to being an employee so he decided to strike out on his own, with a partner on the sales side of the company. The latter preferred Cleveland, then another major armaments city. As my father resigned, the Russian executive who had hired him thanked him for his services, adding that he would never have been taken on had his Jewish origins been known!

The war was at full speed. We moved to the then-edge of the Western Reserve Campus, in Cleveland Heights. Although that sounds pleasant, it was in fact a terrible place; a railroad flat overlooking about twenty rail lines continually carrying freight in and out of town. The flimsy apartment house moved up and down with the freight trains. I was sent to the nearest, if distant, East Technical High School, which had little academic aspiration. Most of its population were kids from the South who didn't really know, or want, to learn anything. We did things like crawl under barbed wire through broken glass in order to get us, I suppose, ready for our role in the war. At this point I was 12. I was again bumped up, this time to 10th grade. I cannot say I learned much, except a little bit of shop, in old East High. All this to say that there are many paths to glory, at least in Theoretical Physics. One of them is enforced ignorance lasting to quite a ripe age. It is really a shame, since those were the years when one could really learn. I did go to the public library often, but was not yet able to make anything out of the things on offer there. It was more an escape from our misery. I mostly read pre-war travel brochures. After a dreary year of near-poverty, and the failure of my father's enterprise, we moved back to Brooklyn.

In New York, my father found as a silent partner a distant relative who had his own watch importing business in Portugal throughout the war and had only come to New York recently. This time we rented a rather dreary apartment not far from Uncle Jack. Our chemical supplies business, in which my mother was also fully engaged, was in a First Avenue loft in Manhattan with a primitive freight elevator and I would often, and enthusiastically, help out on weekends. In due course, they were able to buy a beautiful loft building on East 10th near NYU with my father's partner. [Sadly, the latter got cold feet during the 1970s real estate crisis and sold the property at a loss.] Most cheering was that my parents finally—in the late 1950s—moved to the relative splendor of an airy new apartment after all their years of economic struggle!

I enrolled in my first and only New York High School, New Utrecht, one of the many names the Dutch left around there. The glory of the New York public education system in those days was the plethora of Depression era PhDs who, unable to get college jobs, populated the public school system. They were, and some felt bitterly, overqualified. Still, they were able to instill in us a certain education. I finally made a few friends for the first time, despite the gap in our ages. Noel Corngold, Aaron Galonsky, Billy Proops and I shared a budding interest in science, which extended also a little bit to chess. We also harbored some vague proto-science notions that understanding the world might be fun and could be possible. A good teacher taught us a little bit of early Chemistry and we had decent Math, although just to indicate the difference between then and now: trigonometry was not taught; algebra was the end of the line. I graduated in January 1946. In those days you were guaranteed admission to the City University of New York, including Brooklyn College. At 14, it never entered my mind to do anything else.

There was also a financial advantage. Just as I was graduating, the New York State Regents, the group supervising high schools, initiated something called the Regents' Scholarships. A semi-annual exam was open to all who were about to graduate and the highest scorers (above ~96%) would receive the then princely sum of $350 a year. Tuition at Brooklyn College was free, and there were not many add-ons. As I discovered later, tuition at Harvard in those days was $400 a year (add a couple of 0s to that nowadays). The Regents' Scholarship was reserved for US citizens. I was not yet one, because there was a minimum five year waiting period. By January 1946, I had been in the US four years and about 8 or 9 months, so our principal, likely an antisemite, would keep pulling me out of the exam saying, "You are not a citizen. You are not entitled to take this test". Then the sympathetic vice-principal would send me back in. His argument was that I only had to be a citizen by the time I began using it in the fall. Despite this handicap, I did well enough to be awarded the scholarship. Armed with this victory, we investigated how quickly we could become citizens. I don't know quite what the channels were, but an understanding Federal judge in New York was willing to grant citizenship as soon as the five years had gone by. Since I was too young to acquire it directly, I had to obtain derivative citizenship—through my parents, who therefore had to learn the required Constitutional facts in a hurry. Fortunately they did, although their English was still a little rough. My "derivative citizenship" certificate is still with me.

COLLEGE DAYS

Brooklyn College was a commuting campus. It seemed to me at that time a grand establishment. It was run by a disciple of the prodigy president of the University of Chicago, Robert Maynard Hutchins; ours, Harry Gideonse, attempted to emulate him with the resources at his command. I had to adjust to the fact that college was rather different from high school. I had never taken notes. There is a point in one's education when one begins to get a glimpse of what it means to do science. My first year was extremely instructive, though I only took required courses, including a year of Chemistry. We learned some modern science, although Chemistry has never really been of much direct interest to me. The second year, I took trigonometry, which was not very challenging, but it was the first time I was taught Math at college level. I took Physics 1. In both cases, I made painful discoveries. The first day of Physics 1 lab, we were given a notebook of 8 ½″ × 11″ pages and a fine ruler and told to measure the pages. "They are 8 ½″ × 11″, why should I measure it?" "Because you will see that there are deviations you can map out and see how things fall statistically," the professor replied. To my utter amazement, they were not 8½″ × 11″ uniformly by a couple of decimal places. That taught me something about not taking things for granted, and measuring, and understanding the reason for the measurement. I also had no idea what a proof was. In Introductory Calculus, we were asked to prove something. I said, "What is there to prove? It is obvious." I was raked over the coals and taught that for every epsilon there is a delta. I've never been strong on the higher Math; I've tried, like many theoreticians, to learn as little as absolutely necessary. On the other hand, the idea that there is a proof and a stable domain which follows once the axioms are accepted, was a very deep lesson. Although trigonometry was relatively rote, almost like Euclidean geometry, calculus was a whole new, amazing realm. The year

17

of calculus and then advanced calculus was an eye-opener all around: for example, just the fun of doing definite integrals and the tricks that could be involved. The notion that by thinking you could actually prove something, derive something or solve what seemed insoluble, has remained a wonder ever since. I also took a reading course involving one of Hermann Weyl's books (in German); it proved to be doubly unpleasant. First, because my German was not up to it and I hated to keep looking at dictionaries. Secondly, having looked at them, I discovered that the math was really beyond me. Matrices and groups were something which even poor Heisenberg hadn't learned in 1925 and had to be taught when he came back from his summer on Helgoland, after having discovered Quantum Mechanics. We had good physics courses as well. The House Un-American Activities Committee helped because many young physicists in the 1930s were very idealistic, trained by Oppenheimer between Caltech and Berkeley—a hotbed, in those days of the Depression, of left-wing conviction. Many of them were prevented from having the careers they otherwise deserved. Melba Phillips was one of Oppenheimer's excellent students. Bill Rarita was another; he worked on a paper I was to know very well later, on the Rarita–Schwinger equation. They were all quite dedicated. Despite fairly heavy teaching assignments, they tried to help us. I learned elementary Electromagnetism from Phillips, then Optics. I still think that Physical Optics in its many incarnations was a true breakthrough in what a human being could do and analyze, literally on heaven and earth! Much as I loved it, I only got a B+, as I did not absorb the material well enough. The course was an enormous step upward in modern Physics, an undiluted pleasure, without the burden of having to sweat over Quantum Mechanics (which we hadn't learned yet).

This immediate post-war era was a time of elation, and I could enjoy all that New York City had to offer. In those days, there was no crime, at least in any way that we felt touched by. I loved going to the Museum of Modern Art, where every week silent movies were screened, complete with piano accompaniment. I didn't mind the long ride on the (five cent) subway. For $1 (plus $0.20 tax), one got second balcony seats on Broadway. There was another dimension to the cultural advantages: the library system in New York was amazingly rich. One did not even have to go where the lions rest on Fifth Avenue. The Central Brooklyn Library on

Grand Army Plaza, a beautiful area, was to me a paradise. I discovered all the popular Physics books of the era, written by such masters as Jeans and Eddington, who were then just names to me; E. T. Bell's biographies of mathematicians, etc. The other library section that I found inspiring was on the philosophy of science, which I came upon mostly because it was located next to the science stacks. Since I didn't know any science then, I took it all at face value. I think that Jeans, Eddington and their cohorts, Gamow later, did much to lure credulous young people off the straight and narrow into science, far more than is normally credited to them: they were a recruiting station. I would read books that I didn't understand and I suspect, *a posteriori*, many of the philosophy of science writers didn't understand either. This constituted a parallel education.

So, life at college was a happy set of years. It was a little odd, but not too unpleasant, to be 14, with all the normal-aged high school graduates. I accumulated at least some knowledge and an awful lot more interest. Really, my vocation was formed in this period. You feel it. It is real. It is almost a tangible physical fact at whatever age it hits you. My original interest in science must have started when I was about 12 or 13, when I looked up at the stars and decided that there was nobody up there, so how come? Then what was it all about? In any case, I did not really know at that point that there was such a thing as science that had attempted to systematically answer those questions. When I finally understood, I still didn't know that physics could be a career. Indeed, I had no idea what I was supposed to do with my BS degree. My parents were very supportive, but they left it to me. I spoke to Phillips and some of the other professors, who told me I should go to graduate school to learn real physics. I asked what that was and how one did it. They explained it would take four or five years, and Melba gave me a list of six schools. I had never, or barely, heard of them, including Harvard. Princeton was the only one I knew of because sometime during our college days, Noel and I took a bus just to see where Einstein actually set his mortal feet. We walked up in awe to the Institute. We went by Mercer Street, and even visited the University's Physics department, where we were coldly received by its then-Chair, Henry DeWolf Smyth.

I applied as ordered, and then awaited the thin or thick envelopes, the way students were notified in those days. Noel, who had gone to

Columbia, and I had agreed to meet under a tree halfway between our houses after school on April 15th with our (unopened) envelopes. We each had a heavy one from Harvard, and in my case, a very thin one from Princeton. I didn't take it as badly as Murray Gell-Mann, who held a life-long grudge for having been rejected a year or two earlier. I, on the contrary, am extremely happy to have been rejected because I might have gone there. Their system at that time consisted of courses without grades but with one all-important series of general exams, which would not have been right for me. I also had an offer from a Midwestern school, incredibly generous for those times, like a seven years, no-options Hollywood contract: free tuition, a teaching assistantship etc. So we stood there under the lamp, under that tree, weighing our choices. My Harvard letter of admission was the usual generous Harvard language saying we are willing to accept you, but you will pay normal full tuition the first year, that is $400; if you do well, we'll see what we can do for you the second year. This was still the Harvard of ersatz Yankee homilies, such as "Every tub on its own bottom", meaning as miserly as possible (indeed, my second year's big reward was half-tuition). The only thing I was sure of was that I wanted to do Theoretical Physics. We both opted for Harvard. My final memory of Brooklyn College is, as one of its two Summa Cum Laudes (despite that B+), of leading the class around the track at the graduation ceremonies, thereby thrilling at least my parents.

To my own amazement, after graduation I found myself wanting to celebrate by visiting the Europe we had fled and seemingly left forever. My parents were naturally aghast that I would, at 18, and only four years after the war, want to undertake such a journey. They finally agreed that if I could find free passage across the Atlantic, I could go. After haunting all the maritime offices around Wall Street, I found a Greek Liberty Ship owner who offered me passage if I could be in Newport News, VA in 48 hours, when it was to depart with Marshall Plan coal; of course it took much longer. After two weeks of deciding where to unload, we arrived in still-devastated Le Havre. It was strange to hear French and even stranger to revisit our old apartment in Paris. The concierge claimed never to have heard of us, or our lease, as was to be expected. She then grudgingly conceded that it had been taken over by the neighbors, including the furniture, and could not be seen. Paris was superficially unchanged after nine years and felt surprisingly familiar. I went through Switzerland on my way to Venice, which like the rest of Italy was in dire economic straits but, of course, miraculously beautiful. Locals would congregate at the station, offering rooms for a dollar a night. This was also the case in Florence, my southernmost point. After many stops, my trip ended in Rotterdam, where I discovered the impossibility of hitching a ride back to the US and was forced to take a shabby Dutch student ship to Hoboken, where I arrived with barely subway fare back to Brooklyn. That trip provided a much-needed interlude.

GRADUATE SCHOOL

I'd actually been in Boston once, with my father, and was not especially impressed. It was one of the last post-war cities still stuck in the Depression. It was also racially segregated, with the Irish, Jews, Italians and Blacks each against the other as had always been the case. Then there were the patricians on Beacon Hill, who spoke to nobody. A friend of Noel's father drove us to Boston. In those days there were no Interstates, so our route was US 1 and took something like seven hours, going through every main street in the Eastern corridor. The rooming house where Noel had found a double room while I was in Europe was on the most desirable corner in Cambridge—42 Kirkland Street, just above Harvard Yard. A fellow-denizen was Irwin Shapiro, a bit older than I, but still going strong at Harvard. He taught me driving in his prewar Olds—well enough to pass New York City's nightmare road test. He also took me for my one and only golf excursion, in Far Rockaway, where he lived; it sufficed. Noel and I shared a big room and bathroom. Our sheets were changed, but food was to be eaten out. Jefferson Physics Lab was just a couple of blocks away as one passed the Quonset hut where the meals were served. Harvard required us to pay for meals at the Graduate Mess Hall. This would have destroyed my digestion had I actually had to consume, as well as pay for, them. We would occasionally drop in just to convince ourselves that they were consistently inedible. Harvard in those days was essentially an undergraduate institution in terms of what mattered; the grad schools were peripheral. The undergraduates lived in luxurious Houses with beautiful suites and were served great meals, in expectation of all the donations that were to come when they became CEOs of their family institutions. We graduate students began to dress like them: tweed jackets, chinos, grey flannel pants later on and of course a tie. Furthermore, although Boston's winters are right up there with Montreal's, you were never, repeat never, to wear an overcoat. The most you could do was don a scarf and perhaps some headgear. It was really deadly, but we were young and resilient. We would, on occasion, walk down to the Charles where there were all sorts of athletic facilities, in particular the famous Harvard Sculls. I oared just once; anyone who has ever been in a scull knows that there is less to it than meets the eye. The attendant asked, "Of course, you rowed in prep school?" I replied that, oddly, I hadn't. He said, "Don't worry. It will come easy." He put me in a scull and pushed me off; I continually risked capsizing and then discovered that what seemed like wide streams of water under the arches of the bridges were in fact insidious maelstroms, but somehow I managed to avoid the pillars. After what seemed an infinite interval, I oared back, without smashing the scull against the pier. I swore never again to set foot on any body of water in anything that resembled a scull. I've kept that promise rigorously. Female company was essentially available only at Radcliffe. There were all sorts of rigors about open doors, hours, and I forget all the other custodial words as Harvard considered itself *in loco parentis*. In general, graduate students did not interact much with Cliffies. I think I dated three all told. One of them did open a brief window on the Haut Monde: she asked me to a Christmas reception at one of the better Park Avenue addresses. In the hall stood an enormous crystal bowl overflowing with Malossol, all the rest to scale. That visit removed any aspirations I might have had in that direction! To even get a date involved artifice: I endured a Radcliffe version of *"Our Town"* just to go to the stage door and meet a cast member. Another method was to audit the popular classes—in my time, the lion of Comp Lit's Harry Levin reigned with his *"Proust, Joyce and Mann"* course, in a large auditorium—he would stride in dramatically and hypnotize all the Cliffies from the very first word. It was not a bad course, one that taught me to appreciate those modern heavyweights, if not very productive on the dating front! One last failure: once, coming back from New York on the train, I was getting on swimmingly with a charming Cliffie, until we pulled in and she ran onto the platform, shouting "Daddy" to a full-bore Lieutenant-General. Reader, I didn't have the courage to follow her.

We plunged into courses. The set-up at Harvard was very simple, downright medieval and wonderful in that sense. I had no teaching assistantship, paying full freight, so I could devote myself entirely to study, a

23

good thing. One took four courses a semester for two years, covering the whole required spectrum. We were allowed to take courses in the Math department, which we quickly realized was a deadly mistake because they were meant for mathematicians and were therefore completely inaccessible to us. I signed up for a course from the world specialist on Laplace transforms, thinking it might be useful, but it was mind-bogglingly difficult theorems and extreme cases and I quickly dropped it. I also signed up once for Real Analysis. Little did I know that it was a grown-up game, also impossible. Instead, I took a course in Mathematical Physics within the Physics department taught by Assistant Professor Ivar Stakgold, who had just earned his PhD in Applied Physics, with what was then the record shortest Harvard thesis ever written, 37 pages. That alone made him a prodigy. He taught us from what became *Morse and Feshbach*, one of the standard Mathematical Physics texts for a long time thereafter. It was an extremely good book, especially for us, who had less interest in rigor. However, it had diabolical problems, which we spent most of our time trying to solve. There were also courses from visiting professors; I remember two—very different—ones. One was from Henry Primakoff. He taught us real 1930s and 1940s atomic Physics. We understood what the applications of Electrodynamics were at an elementary, not fancy, quantum level, and it taught us real Physics. You had to calculate, and there were very accurate experimental verifications of the atomic spectra. The other course, its polar opposite, was taught by Stanislaw Ulam. Ulam was a great Polish mathematician who was best known for his profound work at Los Alamos. He broke a couple of the impasses on the design of the H-bomb. Ulam would give us interesting math and recount stories of the Polish Math Cafe in Lwow where mathematicians wrote their conjectures into a special book, where others could provide proofs—or counter-examples. [What befell Polish intellectuals, not just Jews, was total eradication. There was a deeply shameful episode when Heisenberg went to Poland, at the invitation of its butcher "protector", Frank.] Ulam taught us all sorts of very interesting Mathematical Physics that proved instructive in a broader sense. There were also other opportunities at MIT. The distance was measured in walking time. MIT was about 40–45 minutes away, which was nothing in those days, along "Mass Av", Cambridge's main drag, although *in extremis* we would take the bus. The joint theoretical seminar oscillated between the two institutions. We were told to attend it, and that we wouldn't understand a single word. The prediction was right, but we saw all sorts of brilliant people arguing and discussing. This was not quite Soviet-level all-out four hour seminars, but it was still impressive that these great men (there were hardly any women at that time), could have such different views of what seemed like a decidable problem. We obviously didn't understand what frontier research entailed.

We also had less interesting courses, like Thermodynamics and Classical Mechanics; the latter was taught by Philipp Frank. He was Einstein's successor at Prague University way back around 1912. He had written one of the old textbooks, *Frank-Mises*, and would read from yellow pieces of paper that were the book's embryo. He was a very charming Old World personality. Robert Karplus taught a course on Relativity. He was an on-the-make assistant professor at that point. Unfortunately, he did not know, nor did he want to learn, General Relativity. In those days that was almost defensible and widespread. So Bob taught us Special Relativity instead.

The Harvard post-war faculty was highly respected. There was PW Bridgman working on the Physics of high pressures; he was about the only person who did, and got a Nobel Prize for it in due course. He was an old Yankee pragmatist. He had written a letter of recommendation for Oppenheimer, who after graduating from Harvard went off to Europe to learn the latest, in the mid-1920s. He wrote that although Mr. Oppenheimer was Jewish, he fortunately did not have the characteristics of that race and so should prove very pleasant; in those days by no means an unusual stance. Edward Purcell, who (as most of them did) also won the Nobel, taught us Thermodynamics; Norman Ramsey (ditto) second semester Electrodynamics. Wendell Furry, who taught the first half, was one of the House Un-American Activities Committee's victims, and another Oppenheimer alumnus. Harvard had the guts to keep him on, but his career was destroyed by all the political pressure. He was a very smart man and Furry's Theorem was an amazing thing to us as we learned Quantum Electrodynamics (QED), however simple it may appear today. There were a number of others: Kenneth Bainbridge; J. Curry Street in Cosmic Rays; Bob Pound who had many major accomplishments, including measuring the General Relativistic red-shift right in Jefferson Lab. An ever-changing population of assistant professors included Abe Klein, who had gone to Brooklyn College way before me (he had returned from the Army). Roy Glauber, who got his Nobel Prize for pioneering work in Electrodynamics and Quantum Optics, had just returned from a stay in Zurich with Wolfgang Pauli and from there published theoretical papers which underlay much of laser theory and technology. He had been scooped up to Los Alamos as an undergraduate, as had also mathematician Alex Heller, now the Moore instructor at MIT. He was a warm and cultured man, whose mother was a doctor in Greenwich Village; as was Francis Low's, then-Assistant Professor at MIT. The whole first wave of post-war students had passed through; ours came as a letdown. Those first post-war years saw not only Harvard but Chicago turn out a huge number of Nobel Prize winners.

The major first-year required graduate course was Quantum Mechanics (QM); it was usually taught by Julian Schwinger, whom even we lowly peons had begun to revere. He was a child prodigy of the first order. He remained a prodigy, but certainly was not a child anymore, unlike Pauli's old expression that the *wunder* goes and the *kind* stays. He first emerged when City College threatened to expel him for some low grades, so Lloyd Motz took him to Columbia to meet Isidor Rabi for a possible transfer; Schwinger sat quietly while they argued about some arcane QM puzzle. As they glanced his way, he uttered the one word that explained it all— and got him into Columbia. [Rabi became one of the great experimental Nobel Prize-winners and later, a most effective elder statesman. When he went to greet Eisenhower, then the new president of Columbia, Ike said he was always glad to see his new employees, whereupon Rabi replied that it was he, Eisenhower, who was the Faculty's employee.]

Julian was extremely shy and delivered magical lectures without a single erasure or hesitation, while writing ambidextrously. His lectures overawed all of us students: if this was what an average physicist was like, we had no chance whatsoever. Furthermore, his talks were so perfect that all we could do was record them. We all audited his QM course every year, since its content varied. I took it for credit when it was given by Van Vleck and Walter Kohn (another Nobel pair) my first year. The former came from an old (as the name shows) Dutch patroon family that also had the good fortune to own a large part of the Southern Pacific Railway. This must have accounted for the fact that he was legendary in knowing every passenger railway schedule in the United States at a time when there were still railways. Walter Kohn later founded the Institute for Theoretical Physics at Santa Barbara. He was saved from the Holocaust on one of the last *kindertransports* from Vienna. Van Vleck was old-school Quantum Theory, although he knew QM, so he taught us in very graphic language. For example, when talking about interference patterns, he would mention right- and left-moving tigers, whose stripes either enhanced or canceled as they moved across each other; a very useful analogy. Walter Kohn, on the other hand, was of the modern, Schwinger school.

There being absolutely no oversight or organized activities for us, we grad students were thrown on ourselves, which was not such a bad thing given our crop's quality. Many of us have remained friends all this time; an amazing feat of longevity. Irwin Shapiro and Margy Kivelson are still going full speed. A bit later came Sheldon Glashow, who did get a Nobel Prize. There was Ken Johnson, one of the masters of Quantum Field Theory; alas, he died far too early. Noel Corngold went into Nuclear Theory and is Emeritus at Caltech. We went through all the required courses together the first two years. Bob Raphael (who was Jewish) was to become a Carthusian monk, the ones who are not allowed to say anything; not that he ever said much. Bert Malenka had come back from the Army and was TA-ing for Julian, but nevertheless deigned to speak to us. Roger Newton also remained a close friend; Charlie Summerfield, who did one of the last really heavy-duty QED calculations; Jeremy Bernstein, of later New Yorker fame, had a special in: he played tennis with Julian. I want to emphasize the wide distribution of our cohort's backgrounds, a very significant factor. There is no preferred set of parents here; the range is wide: from middle class New York Jews to deepest Minnesota farms, to a rabbi to yellow star war survivors; a few rich, none too poor. That was of course long ago, but shows that love of science is a spontaneous, if rare, phenomenon. An MIT student, David Finkelstein, regaled us—matter-of-factly—about his early days in the Bronx, spent learning symbolic logic while in Junior High. To this end, he decided to read Russell and Whitehead's three-volume masterpiece. The main library in Manhattan could not lend it to him at his early age, but sent him to the Harlem branch,

27

whose copy had never been borrowed. He did, and confirmed having read it all. His later career was mostly in Relativity; he was credited by its true originator, Roger Penrose, with the Eddington–Finkelstein form of the Schwarzschild metric, in keeping with Merton's laws. Robert K. Merton, a renowned sociologist, formalized this all-too-frequent occurrence in his book *On The Shoulders Of Giants* (OTSOG), as Merton's laws:

1. No law is ever named after its true discoverer.
2. That also goes for the first law.

Indeed, even those laws have many other attributions, even if they border on the Russell–Whitehead category paradoxes!

Harvard treated graduate students atrociously, as I have intimated. [Indeed, there occurred a successful, and now historic, proletariat's uprising many years later when the existing students warned all new applicants to stay away.] We were all confined to what had been part of an anechoic acoustic lab. There was a second sub-basement, called Jefferson B-24, reached only, I was going to say, by dropping down a chute. A windowless pit, lit only by neon tubes and a 12 hour clock. In fact, we didn't know a.m. from p.m. There were a bunch of desks. There we each sat and suffered. Occasionally, some of us would break down, so we played all sorts of absurd games, or worse still, tried to solve the *Nation's* nasty crossword puzzles. It was hard to keep one's sanity there. That was when we were writing our theses; it was worse during the years of course work when we had no offices at all. The department had its own library, with all sorts of strange, mostly ancient and uninteresting, but occasionally the opposite, publications—one of which in particular struck me at the time, little did I know why. This was the proceedings of a conference in Warsaw in 1938, sponsored by the League of Nations, just before all hell broke loose, in which the leading physicists of their day presented their work. One of them was by Oskar Klein, the first intimation of something like Yang–Mills theory, if not quite. It struck me as being very important, although I didn't understand why. But at least I was one of the few people in the world who knew it existed. Klein was known for many things, one of them being the Kaluza–Klein five-dimensional theory, and he also got his almost-Yang–Mills out of higher dimensional considerations; that too

28

struck me. I still think that the notion of higher dimensions was a tremendous step. It was invented by Gunnar Nordstrom in 1914; instead of unifying gravitation with Maxwell as Klein had done, he used a scalar gravity model, being pre-1915. That is, his fifth dimension represented gravity as it was then thought be—a scalar field, while the four described Maxwell theory. Nevertheless, it was the leap beyond D=4 that was so daring. Kaluza didn't really do very much, except write down 5D. Klein did a lot more. The jury is still out on how many dimensions we need, but certainly the idea that there could be more than four was truly a giant step. Not all giant steps lead forward, but in any case it was a giant step. So I discovered that in the library, along with a whole bunch of bad leads, as browsing always goes. French Theoretical Physics in the 1930s was in a terrible state because it was governed by de Broglie, a brilliant physicist, who was conned by a couple of "yes" people who were very bad ones, but knew how to cultivate him. They wrote some terrible nonsensical books. Since I read French, I was very proud to get access to this extra wisdom; it didn't take long to realize it wasn't extra wisdom. It was also fun to have a stack card to Harvard's Widener Library, which then had, apart from the Library of Congress, the largest collection in the country. There, I discovered the flip side of stack cards. I went to see Gödel's original 1931 paper in the Vienna *Monatshefte für Mathematik und Physik*. I discovered that someone had cut out with a razor exactly those pages—unforgivable criminal vandalism. The same thing had been done to Planck's four short papers in the *Berliner Berichte*, Proceedings of the Prussian Academy, of 1900. The stack labyrinth led me to open all sorts of books in all sorts of subjects; it was truly a treasure trove. In order to stop student suicides or breakdowns, Widener was specifically closed on Sundays. Lamont Library, which had just been inaugurated, was absolute heaven and the total inverse of Widener. It was a beautiful, airy, building, with all sorts of wonderful literature in many languages, and fancy turntables and records, in those days mono 78s, everything from classics to jazz. The stacks were completely open, if (of course) you were a male. Radcliffe students were not allowed at that time—too much distraction, I suppose. Although it was an undergraduate facility, meant to amuse and cultivate them, one could convince the librarian to let one in. Some of my greatest discoveries, on the literary side, stem from that era. I am forever grateful to Lamont for having these treasures just awaiting discovery. It was there that I completed the arc from Jules Verne to Proust; there was a beautifully bound volume in French lying on a window sill, which I opened to see if French still existed for me: its infinitely long first sentence captured me forever. I have reread it (not just the sentence) frequently.

Perhaps this is the moment to emphasize the importance of literature in my and many other scientists' lives. While we don't usually talk about books, they illuminate our inner dispositions and improve our prose between equations. Speaking for myself, literary discoveries have been almost as exciting as scientific ones.

Harvard continuously ranked where you stood as a student; you could always tell by the way the secretaries greeted you. I alluded earlier to my French Kindergarten's seating method according to perceived weekly merit. That was more or less the way it was at Harvard. Since everybody got an A, just getting one B+ or one A- on some meaningless grounds was easily sufficient to be told, if you wanted to do Theory, that you simply weren't up to it. It was an insane system, perhaps occasionally just, but very much resented. After studying for two years, one had under one's belt a reasonable book-learning level of Physics and were supposed to start research on a thesis. You had to find someone who was willing to be your advisor. Schwinger would take on almost anybody, so I presented myself. I should back up to explain that somewhere during one's second year one had to take *the* qualifying exam. For would-be theoreticians, the exam was in a category called Math and mechanics, basically Mathematical Physics. You would take a couple of months—not off, you still had to take classes—to prepare. We had actually learned much of the material in our courses. I worked like a dog and finally the terrible day came; my committee consisted of Julian plus Bob Karplus and Abe Klein, both of whom were under great stress: they had families and knew they couldn't get tenure at Harvard; assistant professors almost never did. To land desirable positions elsewhere, their job at this point was to convince Schwinger how smart they were. There is no easier way to do that than via an unwitting graduate student. They had just made some nasty calculations in Electrodynamics. In the process they had discovered all sorts of esoteric mathematical functions, in particular, something called dilogarithms, which were still not in the textbooks. They knew about dilogarithms, I did not know about dilogarithms, so this was the moment they could really impress Schwinger, who could appreciate the finer points. During the first hour of that debacle, indeed after 15 minutes, it was clear that I could not contribute anything to that conversation, let alone answer questions. So they proceeded to entertain Julian with their Talmudic knowledge of these new aspects. Not that dilogarithms are very whizzy mathematically, they are just funny integrals that were not commonly known. After I had completely collapsed, they decided I had better get a chance to say something correct, and asked some freshman-level questions, but as anyone will understand, at that point I could not have said how much 2 + 2 was: I was totally destroyed. I was sent out of the room. After a very few minutes, Schwinger came out and said, "You realize that you failed this exam," a rare occurrence in those days. "Yes," I replied. He smiled and I nearly fainted when he added "Don't worry about it." I think that narrow escape was due to my having performed well in a course he had recently taught for the first time, Advanced Electromagnetism. There, he taught us some of the novel work he had done during the war. Instead of going to Los Alamos, Schwinger had opted to stay at the MIT Rad Lab, working on microwaves and waveguides, to which he made enormous contributions, with his knowledge of Green functions. In any case, I was accepted as his thesis student. Julian first gave me the usual little proto-problem, except that this proto-problem turned into a couple of his best-known publications ever. I got essentially nowhere towards what was right; he overlooked that, too. Once accepted, you were supposed to find your own problem, run it by him and then do what you could with it if he agreed—no mean task either. The interaction with Schwinger was hardly one of holding hands, for a variety of reasons. Mostly, he was very shy; he had a couple of people that he particularly liked to have lunch with, which was a ritual during what was supposed to be his weekly office hours. He and his entourage would vanish immediately after class. On one of the few days of the semester when he took pity on us, there would be a queue of petitioners to show him what they had done in the last few months. He would devote at the most half an hour to each of us, pointing out what was nonsense and what might possibly work. These meetings, I should not exaggerate, were extremely useful, if extremely brief. This regime continued my whole third, and more than halfway through my fourth, year,

31

when I screwed up the courage to ask him when he thought I might finish the thesis; "any time you like, even now". I decided to take him up on it lest he change his mind. The whole PhD thesis process involved a very tight set of specifications. You had to use a kind of Bible paper that was only available in one or two stationery stores. You had to present three copies, each pristine. In particular, all the equations had to be written in by hand by the author. Since my writing would be the envy of any MD, and no erasures were permitted, this was just one more ordeal. They had to be bound, with gold backing, and presented by a certain institutional deadline, that I just made—a great relief.

Writing a thesis, depends on one's intellectual level and professor's expectation, and more than anything else, on one's scientific maturity. It is supposed to produce, literally as in the medieval guilds, a "masterwork": an original project that proved you had achieved "mastery" of the field. That has always been the point of a PhD. My thesis was by no means my best, work, but it was a beginning and got me out. It was the equivalent of going from apprentice to journeyman. There are many other factors not always under one's control. For example, there are rather dry periods and then there are periods, as in 1925, that the old-timers said all you had to do was sit down and write anything and it would become a classic paper, because QM had conquered the world. My period, and that of my cohorts, was, by and large, one of transition: the great post-war victories were in Quantum Electrodynamics, of which Schwinger was one of the leading practitioners. Schwinger, Feynman and Tomonaga, with the help of Dyson, had more or less understood and explained how to ask questions in the subject and get (mostly) sensible answers. By my time, everything possible had been done that didn't require modern computational tools. Even when I had achieved a PhD and done a lot of calculations, I didn't really understand the then-frontier in Elementary Particles at a level that would enable me to do interesting work, or even make useful conjectures. Indeed, I only began to understand Physics (that is, when I could do original work) after my four postdoc years. In fact, the process of education never ends. Hence the idea of the old apprentice system, preferably far from one's original institution, so as to get a wider perspective of what kind of Physics is being done, by whom, where and in what style.

Scientists in earlier times had to have independent means or other work, because they were not paid for being scientists. Except for the odd astronomer, this was true at least until universities much later included Mathematics and the physical sciences. Lavoisier in France, who lost his head during the Revolution, was a tax collector. Michael Faraday was a bookbinder's apprentice. Indeed, to embark on a scientific career before the mid-1930s, certainly in the US, was a very dicey story as there were very few positions. Families had to have the leisure and the funds to support their children with no very clear career outcomes. The elite high schools in New York, Stuyvesant, Bronx Science and others, were then a source of future PhDs in the 1930s and the 1940s. The refugees from Europe also contributed. Nowadays, it is quite different because graduate students expect to get not only free tuition, but enough to live on. There are a great many more departments issuing PhDs, with larger staffs, than in our day. That's a big difference and makes it much more democratic as to who can dream and be entitled to do science—a very good thing.

I was lucky because my parents were able to pay the tuition the first couple of years and cover my modest living expenses. Boston at the time was not a place that was tempting for large-scale expenditures. We lived monastically. For my third year, I moved to an apartment with the late Dick Arnowitt, with whom I was to do some of my best work at the beginning of our careers, and Bob Raphael. It was pleasant, also located in the heart of Cambridge. We tried to lead a normal life, even occasionally using the kitchen. Indeed, there occurred a notorious incident when we dared invite the Schwingers for dinner. We sat them at the head of the table when, a few minutes later, there was a crash and Julian's rickety chair collapsed. In a brilliant save, we managed to get him onto another one before he even realized what had happened.

We also took long excursions, mostly on foot, to such meccas as Durgin Park, the beanery par excellence. The Commonwealth (never the State) of Massachusetts began to impose taxes on meals over $1, which impelled Durgin Park and others to offer 99-cent lunches to which we grad students were (naturally) drawn. The walk revealed a cross-section of the city, as one came from MIT with Beacon Hill on one side and "the General" (Mass General Hospital) on the other, then up to Scollay Square, the dispirited center of the red-light district. This led to the financial and historical parts, with Faneuil Hall (far from its later restored glory) and then a hike up some grungy steps to Durgin Park, in no way to be mistaken for a decent restaurant. My parents were treated to both my MA and PhD ceremonies, complete with crimson robes and handshakes with the President. [The MA was a formality, just requiring a $10 payment.] I suppose—though we never discussed it explicitly—that it was for them the culmination of the immigrant's dream for their progeny, or at least the assurance that I was launched on the vocation of my choice. I cannot say that those emotions were also in my mind—just surviving the four-year ordeal was reward enough.

For those of us getting our PhDs at that time, it was unlike now, when everything is on the web and all is automated for recommendations and applications. Schwinger had a unique understanding with Oppenheimer, then Director of the Institute for Advanced Study (IAS): he would simply send Oppie his relevant graduating class, which usually consisted of one or two people, and IAS would take them. Unfortunately for Roger Newton and me, we were following a tough act: while at IAS, our immediate predecessor had snapped and been caught climbing a local estate's wall at night. It was all handled properly and he was airbrushed from all records. [I was told, perhaps facetiously, that he became a psychoanalyst.] As a result, Oppenheimer decided he'd better vet us, so we arrived at his office, at the time guarded by FBI agents because of his hyper-sensitive files— before the notorious spring 1954 Hearings. He asked me for the title of my thesis and proceeded to tell me its contents and why it was wrong; the latter was possible, but not the former—he was wildly off. Roger was similarly inspected, and passed the "tic" test, for we were accepted. In those days, with a little luck, a postdoc got something like $3,200 a year. Even then, this was not an impressive sum, certainly not in a city as wealthy as Princeton. It had rich undergraduates, all male. We were not at the University, so we had nothing to do with them, but it set the tone for the town, along with all the bankers and lawyers commuting to New York.

Before beginning this next phase, I decided to spend the summer of 1953 in Europe. It was in considerably better shape than in 1949. The Les Houches Physics Summer School had recently been established, so I spent a few days there. Against my will, I was (literally) roped into mountain climbing. I should not call it an expedition, but for someone who had never gotten higher than New Hampshire's Mt. Washington on foot, this was quite a tricky undertaking. Indeed, we went quite high, led by Bryce DeWitt. His wife Cecile ran the Summer School. She went on strike at some point and said she wasn't climbing any higher, pretending a new pregnancy as her excuse. I happily said we couldn't just leave her alone so I would be chivalrous and give up the rest of my climb. We sat on a tiny ledge for several hours freezing and attempting not to look down, while trading Physics problems and questions until finally the rest of our party reappeared. We made a grateful retreat back to Les Houches. I met quite a few people there, one being Tullio Regge, who died recently. He was still a graduate student and seemed to me destined for great things, which was not difficult to guess. My trip encompassed quite a stretch of territory. It included a visit to my uncle Emile, the driver in our hair-raising escape from Paris, who will be recalled to have ended up in Israel. I rode 4th class by train through the Balkans to Athens, then took one of the first El Al flights on an ex-military transport. Ours was a tearful reunion, including meeting his welcoming wife and children. He couldn't do enough for me. I remember being sent off on a flight to Eilat and getting out of a DC-3 onto an airstrip in the middle of nowhere: Eilat was not yet a city. I walked down to the beach to try and get into water as soon as possible since the temperature was truly unbearable. The sea did look very attractive, except that leeches proliferated around its edges. I decided to risk death by bleeding rather than heat stroke, so I jumped in, then took the next possible plane back to Tel Aviv. Many other adventures followed, including deck passage from Istanbul to Naples (shades of the Aenead!), leading to the wonders of the Amalfi coast before mass tourism took over, and eventually a final ship home.

Over the decades, I visited Israel and its extremely strong Physics community, watching its transformation into a first world country at such places as the Weizmann Institute, Tel Aviv, Jerusalem and Haifa. Indeed, I collaborated fruitfully with Adam Schwimmer, who once booked me into Weizmann's then new guest house's royal suite, complete with a hand-decorated, gilt, toilet bowl—not a normal experience in our trade! Israel is an informal country, so I was surprised when an old colleague, Haim Harari, then head of the Weizmann, greeted me in a tie and jacket. He looked uncomfortable, but explained he had to see the country's

President, who would be equally ill at ease in his! Yuval Ne'eman was perhaps the best-known Israeli physicist; while a military attaché at their UK Embassy, he pursued a degree with Salam, sharing a major discovery with Gell-Mann while doing so. He came from old Sabra stock and was a staunch expansionist. He was even elected to the Knesset at some point and took me along to its cafeteria and a session where a fiery Meir Kahane, soon to be assassinated, ranted on. He also invited me to dinner with Zubin Mehta, a rather unexpected experience in the higher realms.

THE INSTITUTE FOR ADVANCED STUDY

Come September, it was time to become a Jerseyite. I rented the chauffeur's quarters above an estate's garage, to begin my stay at what was then, and remains, one of the world's preeminent institutions—one which has hosted just about every great, and many of us not so great, theoretical physicist. Nowadays, it houses a much wider collection of fields. I was given a little office in a small brick building poetically called "D". Right next to mine was Lee and Yang's office. Out of it, screaming could be heard all day as they argued. They were in their Statistical Mechanics phase, a distance still from Parity violation. Mandarin screaming on any topic couldn't help me much. I had little to do with Dyson and Pais either. Nobody was interested in us young postdocs. Pais did do his bit to be slightly sociable to me after hours—once. I knew that the war had caused him suffering back in Holland, but he never really talked about that. Much later, he wrote his memoir on that period, in the Hudson Review. It is harrowing reading to this day.

The year at IAS traditionally started off with a high-octane cocktail party at the Director's house, supervised by Mrs. Oppenheimer, a legendary Martini expert. Being new, I did not know when to take my leave until Valentine (Valya) Bargmann, an older mathematical physicist (and professional-level pianist) took me firmly under the arm and said "Sabbath is over." By the time I saw Oppenheimer's secretary arrive at IAS in her (vintage, of course) Rolls-Royce, I knew this wasn't Kansas! There were always impressive visitors such as Gian-Carlo Wick, who had left Berkeley rather than sign their disgraceful loyalty oath during the Red scare. He had been Bruno Zumino's mentor in Rome and was at the height of his career, yet approached us newcomers as if we were equals. The great Yoichiro Nambu, as yet unappreciated in the West, came from Japan.

I was a bit uncertain as to what to start work on. Luckily, two people had designs on what they thought was my expertise. One was Marvin (Murph) Goldberger, an assistant professor at the University of Chicago, who was spending a year at Princeton. He eventually returned there and became head of its department. [Instead, some three decades later, I was invited to Princeton as visiting professor for a short—and very productive—stay.] His later distinguished career included serving as director of the Institute, and president of Caltech. The other was Walter Thirring, who has written his own memoirs; both died not so long ago. As a young assistant professor, Walter was temporarily at Bern University. His father was a famous physicist who headed Theoretical Physics at the University of Vienna. Walter had to go abroad and prove himself, which he more than did. He eventually came back and took over the mantle, of which he acquitted himself at a very high level indeed. They came into my office one day and said, "You are one of those Schwinger prodigies and you know all those new tricks of his. We wondered if they could be of use in a new way." They were interested in the functional methods that Julian worked magic with, so I modestly agreed to try. This was not false modesty, since I knew only a little about it. We set to work and tried to do some low energy theorems for the then meson theory of strong interactions. We wrote a couple of papers and I learned a lot more than they did. They were both married with children, so they would go home at 5 p.m. I would stay up late into the night. [There wasn't much of a temptation to go out on the town.] The next morning, somehow, despite their domestic duties, they had usually gotten further than I—certainly never behind—anything that I had calculated during my long night. I began to realize that talent was a very useful thing to have. It was a lot of fun to do adult research at last. As we finished, the interest in dispersion relations, that other approach to elementary particle theory, was just beginning. Murph thought that would be really promising and wondered if we were interested in dipping into it. I decided, brilliantly, that it was not for me. This mistake was remedied by work I did a couple of years later.

Two quite disparate events in my personal life occurred during that year. Most momentous, I met my future wife, Elsbeth, at a dinner chez Goldberger. I need not expatiate on the impact of love at first sight, especially when it is the start of a lifelong relationship. Indeed, its providential nature was emphasized by the fact that Elsbeth was only visiting Princeton briefly, on her way back to Scandinavia from a course at Berkeley; her stay, I learned, was due to her father's, Oskar Klein, connections there, made during his own recent stint at the IAS. While not herself a physicist, she was of course very familiar with the milieu, which made it even more fun for us to share many future experiences together.

The other event might have had a fatal impact on my career. In late spring 1954, the Oppenheimer hearings stripped him of his clearance so the FBI and the filing cabinets were all gone. Everyone at the Institute was naturally incensed about this and the role his nemesis, Edward Teller, who ran the US atomic bomb program at that time, played in the inquisition, which was really a struggle over our military policy and whether to build the H-bomb. The wartime draft was still on, although the Korean war was finally over. My local draft board, located where my parents lived in Brooklyn, decided that they had had enough of my deferments, and decided to end my soft life while it was still possible. Since I was 1-A, the top eligibility status, I was told to appear for medical exams. This hit me just as I was beginning to be able to think and work—a two-year interruption would kill my budding career. So I went to Oppenheimer, though I assumed he no longer had any power after the Hearings. Instead, he said, "That's no problem. We can fix it." As I sat in his office, he told his secretary on the intercom, "Get me Teller." In those days, long distance was still a big deal. I thought to myself, "How could he still talk to Teller?" I overheard only his end of the conversation, of course. "Hello, Edward. How are you? Everything okay?" Then he said, "Yes. Fine here too," which sounded alien. Then Oppenheimer said, "We have a little problem," and described my circumstances. The next thing I heard was his saying "Edward, I knew you'd be able to fix it. Thank you very much. Goodbye." He hung up and said to me, "It's all set. You are going to report to work at the Rad Lab in Berkeley in a couple of weeks. You'll need Q clearance, but I'm sure that will work out." Over the next couple of weeks, I heard from my parents and various relatives that FBI people had questioned them. Apparently the agents were satisfied, because I was soon told I had the very clearance that Oppenheimer had just lost. On arriving in Berkeley, I was sent up to Teller's office, where again I couldn't believe my ears. Teller said, "Welcome to the lab. Tell me, how is Oppie doing these days?" I muttered whatever I could. He said, "That's good. I am having a little gathering in my house tonight, at which I will also play some piano. Please come if you can." It was a soirée in the European style. Teller played on a concert grand, as befitted his professional-level renditions. I never saw him again, but better, I never saw my draft board again, either. My exemption carried me through to the end of the draft a few months later. Another miraculous save by an evil spirit, as we thought of Teller in those days.

The summer at Berkeley was spent under armed escort in an auditorium where we were lectured on Statistical Mechanics. It turned out that we were attempting to get fusion for civilian purposes. It was called Project Sherwood, although even the name was classified. On that score, shortly after I came back to Princeton, Eisenhower decided to make a gesture to turn swords into ploughshares by telling the Russians what had been learned at Project Sherwood. That turned out to be embarrassing, since the Russians knew far more about peaceful fusion, and how one might get it to be useful, than the US did. It did help cement some sort of summit meeting, albeit one marred by the U-2 incident.

During this West Coast summer, Dick Arnowitt and I published a paper with a strange background. Marcel Schein, an expert on cosmic rays, had just observed an event that was completely off-scale in terms of energetics compared to any other that had ever been previously seen. There was clearly no quick explanation for it, and he never took back his results. So Dick and I thought very hard and discovered that we could gain a factor of 10^4 or more by assuming that this event, unlike normal electromagnetic showers, was due to a magnetic monopole. Magnetic monopole charges, as Dirac first brilliantly showed, must have strength inverse to electric charges, but would generate similar electromagnetic fields and radiation. So we came up with at least a plausibility argument, which fell on all-too-fertile grounds in Berkeley. Luis Alvarez, one of the great experimentalists there, was also involved with the Air Force. At that time, the Strategic Air Command (SAC) was patrolling near the North Pole. When Alvarez heard about us, he suggested that we take a ride in one of the Air Force B-29s and see if we could learn any more. Presumably magnetic monopoles would have some sort of affinity with the magnetic North Pole. Although we were fond of travel, the idea of going up in a plane laden with H-bombs did not seem too enticing. It was clearly Dr. Strangelove ahead of its time, so we politely declined. That shower has never been refuted—or confirmed. Many years later, Dyson proposed a more prosaic possible explanation, and there things stand.

Physics also leads to non-academic adventures. In 1952, at the end of my third graduate year, a very smart young Welshman on a Commonwealth fellowship from Cambridge University appeared. Sam Edwards knew Julian's formalism backwards and forwards, and was particularly interested in applying it to Condensed Matter Physics rather than relativistic theory. In this, he succeeded spectacularly; his other services to UK Physics earned him a Knighthood as well. Sam asked me whether I'd be interested in joining him and a friend from Harvard Law School. As the latter was already in Mexico with his car, we would have to get there on our own. I jumped at the chance to have a restful break. Sam and I spent several days on a Greyhound at the height of summer, from New York through the south to the border city of Laredo. We got some faint idea of segregation on that leg. Every drinking fountain, at every miserable bus stop, had at least two copies and many variations of restrooms for men, women, and Blacks. We caught another bus in Nuevo Laredo, on the other side of the Rio Grande. At that time Mexico City was not the present day 30 million megacity. It had only about two million, a spread-out, almost rural, sort of place. There were no highways; then again there was no traffic. You parked anywhere. On every block there was a man you paid a peso or two to (symbolically) watch your car. We saw many sites, including a trip to Acapulco, which at that time was a grand sort of place. We got a room at one of the skyscraper hotels. The temperature and the tides were ideal for swimming. I went way, way out. On a hidden rock underwater was a sea urchin, which shot a huge number of poisonous spikes into one of my heels. I somehow got back to the beach where I was immediately given a generous spray of tequila. Unfortunately, it was administered externally. A doctor removed as many spikes as he could. There were a few bad days, but those were quickly forgotten as we toured that marvelous country. There were no drug gangs then. Perhaps the most dramatic trip was up one of the two fiery volcanoes above Mexico City, which had a sort of trail navigable to a 1952 Ford.

I had a similar adventure in my first year at the Institute, with John Ward and Dennis Sciama. John was a great physicist, known for Ward's

Identity (indeed, he was the most unsung hero of QED—and even of the UK H-bomb!). With Abdus Salam, he also came close to understanding the Weak Interaction part of the Standard Model well before its time. John had full-scale Oxonian manners and accent. Dennis, equally Oxbridge, did original Cosmology and almost single-handedly built up the UK's formidable standing there. One of his early students was Hawking; on a visit, I asked Dennis if he had any decent students lately. "Yes, I have one who's quite promising, but I can never pin him down—if only someone would nail him to a chair…" Oppenheimer had invited Dennis to IAS to show his dislike for General Relativity; it was not much fun for him. We three decided we were stir-crazy and had to get out for a break. We would drive from Princeton to Miami and back, over the long Thanksgiving weekend. There was no Interstate and we went through every speed trap on US 1 through South Carolina and Georgia (the southbound tourists were their revenue source). In Miami Beach, we went to the smartest place in town, the Fontainebleau. It still exists. It was already a self-parody, a sort of gold-plated Borscht-belt place. It had a beach and the water was great. Unlike that time in New Jersey, which was very grey and not very pleasant, the sun was warm. We were the only ones not at the pool, but at the beach. My cohorts decided this was too good to leave so they stayed a little longer. Unfortunately, I had to get back, since I was working with Goldberger and Thirring.

Soon after, John sent me a postcard from Lebanon on his way to the Far East with Salam. The front of the card was the racy sort of image you could buy in Beirut. The back said, "We have finally discovered the secret of the Weak Interactions. More later." He finally ended up in Australia, alas without sharing the Nobel he deserved. I also met Benoit Mandelbrot at IAS, then an Assistant Professor at the University of Geneva. I don't have to give details, since he wrote an autobiography, alas published post-humously. He would frequently call me over the years—especially when he heard interesting Physics news; this also helped me keep up my spoken French.

Postdocs did not give seminars at the Institute. They would have been destroyed. Indeed, Pauli frequently destroyed bigger fish when he was visiting. He would tear people apart so thoroughly that they would leave the seminar room in tears. Pauli was a very great man, but totally devoid of the usual polite instincts. I discovered recently among some old letters, one from him. When I was going to Copenhagen, I thought it might be instructive to spend a short period in Zurich. I wrote Pauli, whose reply was, "Unfortunately, I am not the Swiss Consul and therefore I cannot deny you a visa. If you come, we will just have to accept that fact." That stopped me cold. Later, I was told that this was an enthusiastic invitation by Pauli's standards.

My second year at the Institute began auspiciously. Murph and I started working on an interesting, experiment-related and timely question. As Elementary Quantum Mechanics tells us, electrons in the lowest state of hydrogen atoms spend most of their time at the nucleus. As it was becoming possible to make pi-mesic atoms, with negatively charged pions replacing the electrons, we realized they would yield otherwise unavailable details of the strong pion-proton interaction from the shift of those lowest states relative to their (electronic) Bohr levels. We worked it all out, then discovered that Thirring and one of his students back in Europe had done the same thing at the same time; we decided to write a joint paper. I have never met the fourth author, Baumann, and I think he never met Murph, either. Before submitting, Murph insisted that he talk to Fermi in Chicago and get his blessing before we published. You could hear the deference in everybody's voice when Fermi was mentioned: he was known as the Pope, because he was infallible. About thirty years later, I was giving a colloquium at Neuchatel and someone asked if I was related to one of the authors of the four-way paper. He showed me a tattered reprint and explained that he was part of a group that was continually improving on experiments of this type. He said that everybody in his group had a copy of that paper memorized and held it in their pocket at all times. I should have written another such experimentally-oriented paper, but this was unfortunately the only one.

That same year I got the chance to hear one of Einstein's last seminars. Oppenheimer (of Oppenheimer-Snyder black hole fame, ironically his one claim to Nobel glory) had gathered us new recruits to warn against having anything to do with the "old fool down the hall" or with Relativity in any form. There was little danger at that point, since none of us even knew what GR was. Little did I know the central role it would play in my later work, or indeed in current Theoretical Physics. Indeed, I was even

43

asked to write the article on GR for the French successor to the 300-year old Encyclopédie, as well as its update, in collaboration with my old friend Thibault Damour, thus paralleling Eddington's in the Britannica. It was nearly impossible to discover when and where Einstein would talk, an apparently classified datum. One day I got a tip and went to an unmarked seminar room where Einstein held forth in German with his latest unified theory. [I have always wondered how he would have reacted to the elegant unification so simply achieved by (Quantum, yet) Supergravity just two decades later!] Einstein was to die that April. Sadly, the first international Relativity meeting ever (hence called GR0) was scheduled for June in Bern to celebrate fifty years of Special Relativity as well as all the other great Einstein achievements of that miraculous year of 1905, and the 1915 birth of General Relativity.

My other Einstein experience was sharing the Institute van with him and Kurt Gödel. One terribly rainy day, my usual walk to the Institute necessarily turned into a van ride. After I was picked up from my apartment, it stopped at Einstein's house, whence he emerged wearing his uniform: a sailor cap and a sweatsuit. I was cowering in the rear, as I was to do the entire journey. He sat down in the front and took no notice of me. Shortly after, the van stopped again, this time to admit Gödel, who was wearing just the opposite gear, that of a Viennese Herr Professor, a Homburg hat and a black suit and tie. He was a tiny, thin figure; he paid no attention to me either, but doffed his hat and greeted Einstein in German. Einstein doffed his sailor cap. Alas, no profound gems were exchanged during that short trip. On arrival, they again doffed their headgear and vanished. One famous story about them: when Gödel was to acquire American citizenship, Einstein was his sponsor. Gödel read the Constitution and (of course) discovered some logical inconsistencies (Article V as an obvious example of category error). When the judge asked Gödel if he had read and understood it, he replied, "yes, but...", at which point Einstein kicked him sharply under the table, reminding him not to pursue it.

John von Neumann was another of the IAS greats—always dressed like an old-world banker; his main interest then was the ENIAC, the first large, still tube-based, computer, housed in a wooden hut. Jule Charney, one of the fathers of modern climatology, was an early user. Andre Weil, another legend, with a colorful past and an even more famous sister, Simone, was occasionally visible; he and the other mathematicians were always engaged in some vendetta with Oppenheimer. Unlike now, there was very little interaction between our disciplines. Indeed, I am reminded of the old Hilbert story. He was to give a special lecture at some nearby institution but being absent-minded, he asked his hosts, as he was about to begin, to remind him of his subject—"Mathematics and Physics." He paced across the stage, mumbling "Mathematics and Physics", "Physics and Mathematics"... Finally, stopping mid-stage, he exclaimed "they have nothing to do with each other!" and sat down. This from the father of Hilbert space, the Einstein–Hilbert action and much else!

As I mentioned, the Institute seminars could be very rough, depending on who was in attendance. Pauli once attacked Pais who was talking on his work with Gary Feinberg of Columbia. Pauli made very clear he did not like it and essentially stopped Pais from speaking. Pais finally said in German, "Pauli, Ich kann viel."—"I know a lot." Pauli replied, "Ich kann mehr."—"I know more." That was the end of the seminar. Pauli was also violently opposed to Yang–Mills theory, which he had apparently discovered earlier, but rejected because there was no room for massless vector mesons at that point. Fortunately, that did not stop the theory from propagating. This was the final stage of its creation after Klein's 1938 higher dimension inspired version, which was not quite the right one. Pauli was not unique in inspiring awe: Bohr and Dirac also visited, which, to us postdocs, was like being transported to Olympus, and rightly. True to legend, Dirac often walked in the woods and liked to do some chopping. Neither was very active at the seminars, though Bohr would sometimes utter some deep, if inaudible, remark; if he said "very interesting", you knew he thought it was nonsense. That Dirac was our deity was not surprising. He was also Feynman's, who was hardly a hero-worshipper: many years later, at the first Coral Gables conference, I heard him ask, as Dirac entered, "Professor Dirac, how did it feel to discover that equation?" with the wonder it inspired in us all.

Oppenheimer felt that it was his responsibility to launch us postdocs on our careers. He was consulted by all physics departments in the country, then in full-scale expansion. Oppenheimer had decided that my next step should be an assistant professorship at a large Midwestern university.

I protested that I needed another postdoc stage to develop further, though in those days two years were deemed sufficient. Four years was exorbitant and beyond that was simply unthinkable, very different from now. In particular, I wanted to go to the Niels Bohr Institute, a mecca since the early 1920s and join my future wife, who had returned to Denmark. He said that even if I didn't want the job, I should still visit the school in question. Of course, I obeyed. After my talk, I was ushered in to see the head of the department. There were presidents-for-life in many places, having the full power as Chairs to do whatever they deemed essential to their growth. I was immediately offered the assistant professorship, based on Oppie's say-so. I tried to disentangle myself gracefully and finally the Chair said to me, "Then what would get you to come here?" I was thrown into a panic since actually nothing would, but having to say something, I replied, "I suppose an associate professorship." I figured that might do it and indeed it did. After a frosty silence, I was ushered out.

THE BOHR INSTITUTE

Luckily, the National Science Foundation (NSF) had recently endowed some postdoctoral fellowships, for which I was eligible. I applied and was awarded a two-year stay at the Bohr institute. It was customary to be interviewed by Niels Bohr, an historic occasion for anyone. He was fiddling with his pipe, unable or at least unwilling, to succeed in lighting it. After a few pleasantries, he asked me if I had any questions. I said, "Professor Bohr, I do have one question about Quantum Mechanics and Measurement theory. Quantum Mechanics is supposed to be the kinematics of Physics, that is to accommodate whatever dynamical theories were involved. Yet the existence of measuring apparatus, let alone observers, required there to be macroscopic objects that were themselves effectively classical. That's how the transition to macroPhysics would arise. On the other hand, for there to exist macro-scale apparatus, long-range forces are required, such as Maxwell's theory provides. Without those specific dynamics, there can be no true QM." His efforts with the pipe redoubled, he coughed discreetly and I realized that was the end of the interview.

I produced only one paper during that time. It had an interesting background. This was the very beginning of interest in Quantum Gravity. In particular, hope had been expressed by a number of great scientists— Pauli, Landau and Klein, independently, that Quantum Gravity might provide a cutoff to the ultraviolet problems of Quantum Field Theory. Theirs were just hand-waving arguments. So I decided to see if I could wave a little bit less and came up with an explicit one. It was published in *Reviews of Modern Physics* as part of the proceedings of the January 1957 meeting in Chapel Hill. Hosted by Bryce and Cécile Dewitt, it attracted a respectable number of people, including Feynman. When I presented this work, Feynman went into high gear, tearing down what little evidence I might have presented. He was quite correct. At that point it reduced me, if not to tears, to despair. This was my second year in my second and last postdoc and I had no job, nor any particular prospects of coming back to the US. To add to my travails, I had taken the long propeller trip from Copenhagen for the meeting and managed to catch the worst flu I had ever had in my life, including high fever. One of the most assiduous visitors to my hotel room was Feynman. He must have felt that he had precipitated my virus. Although my conjecture was no wilder than other people's, I realized that I had not really thought it through enough.

I still had not reached a full understanding of what Physics was. I felt pretty keenly that I wasn't ready to do anything interesting. I began to understand why the motto on the facade of the Royal Theatre said "Not just for Pleasure."

While my research was rather dispiriting, my personal life was much more successful. In August 1956, Elsbeth and I were married in Copenhagen with a wedding dinner at Tivoli Gardens attended by both sets of parents and our closest Danish friends. This was the beginning of a lifelong friendship with her parents, Oskar and Gerda Klein, whom we often visited in Stockholm. Oskar and I had frequent Physics discussions along with his best student, the late Bertil Laurent, given our close interests. Despite his brilliant achievements, Oskar was a very modest person so I never felt intimidated. He was still active when we met; he gave the Nishina (of the celebrated Klein–Nishina formula) lectures in Japan, and I was able to bring him to Brandeis later for a successful term. An amusing fact about Oskar is that because he was clearly a *Wunderkind*, he was taken into Svante Arrhenius's lab at 16, where he worked on real Thermodynamics. The lab was at the large family property outside Stockholm. In due course, a grandson Erik needed home tutoring, for which they turned to Jan Klein, Oskar's eldest. This in turn led to Erik's meeting of—and subsequent marriage to—Birgit Klein, my wife's youngest sister, thus allying two scientific dynasties.

An American postdoc abroad faced a difficult re-entry problem. In those days especially, you were quickly forgotten in the US. Not that I was known before I went, but at least I had Oppie and Schwinger. Fortunately, Julian offered me a lifeline, to spend a year as his assistant and lecturer at Harvard. I would have to teach a course, but I would have time to think, get work done and be repatriated. I embraced that chance.

I still have the 1955 Einstein meeting Proceedings, including a list of attendees, whose total number was about eighty, a large fraction of whom were there for ceremonial reasons. There were precisely four of us who were under a very high age. The rest were people whom I thought of as relics from the past, like Fokker and Weyl (who had given some brilliant Math lectures at IAS while I was there), and Fock, one of the rare scientists then allowed out of the Soviet Union. He was an ardent Party member. He wore, as befitted his status, the latest Russian hearing aid, the size of a large suitcase. For those of us sitting in the last row, there was quite an interesting array of great men. The younger crowd included Wally Gilbert, then a graduate student of Salam's in Cambridge. He had just come from Italy, so he wore one of the then-fashionable Italian men's silk suits. I greeted him with one of my few words of bad Italian and he answered likewise, until we realized that it was not our common native language. There was Felix Pirani with whom I was to become close, personally and scientifically, and John Moffat who also came from Cambridge at this point. We four were the younger generation. Neither Wally nor I knew anything about Relativity. Felix was already promising in that very small field. He was at Kings, where Bondi was establishing his school. Moffat also leaned in that direction, although he had worked with Dirac on other matters. This was the first time many of the old heroes had ever met each other. Little did those early practitioners expect its explosive growth not so long after.

One important event, late in my stay, was the founding of the CERN Theory division in Copenhagen, since its true seat, near Geneva, would take a few years to complete and the creation of a temporary base was a morale-building gesture uniting the sponsoring countries. Many younger people arrived, including Bernard Jouvet, a brilliant skeptical French particle physicist who died too young and André Petermann who stayed on at CERN for life, though mainly invisibly, showing up only in the wee hours. His accomplishments were many, if mostly unsung: only one, the renormalization group creation in his thesis is well-known. His advisor was the aristocratic Baron Stueckelberg, always accompanied by his large dog at (later) CERN seminars; both were tragically under-rated. They were antipodal: André came from the slums of Geneva (yes, even it used to have some—albeit higher class) with an accent to fit. He also invented quarks independently of Gell-Mann and of Zweig, did still-not-widely-known important calculations including in QCD... and raced cars. There was also a Belgian, Marcel Demeur, whom I next saw some 30 years later as I passed his office door on my way to give a lecture in Louvain. When I said hello, he responded with *"Quoi de neuf?"* Alas, the new group started too late to be of avail to me.

One final escapade from those Bohr years happened in 1955, as three of us Physics postdocs and one friend from Harvard Law School piled into a car for two weeks, intent on eating at every three- and possibly two-star restaurant in France as rated by the Red Guide Michelin. Paul Martin, Irwin Oppenheim, John Kaplan and I would atone for this gluttony by looking at every three-star site recommended by the Green Guide Michelin. At that time, classic French cuisine ruled. It's hard to pick out any one favorite but the restaurant in Avallon in the heart of Burgundy stands out, if only for its wines. Indeed, sometimes we would eat one feast for lunch and another for dinner, driving hours in between. To our amazement, it worked out, without our dying on the roads or becoming sworn enemies or getting gout. We could—barely—cover the restaurant bills, though we slept in tawdry holes in an insane attempt to balance our budget. Most surprisingly, this degenerate escapade did not harm our careers!

AMERICA AGAIN

In late summer 1957, my wife, our newborn daughter Toni and I embarked on the Bergensfjord, a Norwegian liner. It was a lucky summer crossing with beautiful weather. We were met at the pier (a genuine West Side one this time) by my parents, who were so happy to lay eyes on their first grandchild. After spending a few days with them, we set off to Cambridge to find an apartment. We rented one in a Boston archetype, the triple-decker. It was, to put it mildly, a stark contrast to Denmark's smart apartments, but very convenient: I could roll out of bed and go to the office by crossing the Harvard Law School lawn. We shopped at the local A&P and First National on Mass. Ave. These were still the old-line depression days in Boston, the last city to adopt new trends. The "supermarket" was only super to the extent that there was a long counter. After the purchase was complete, the clerk would tote it up, for all I knew, on an abacus.

I had never taught before; my assignment was Classical Mechanics, which was supposed to be easy. Unfortunately, the class I faced was a bunch of smart graduate students, many of whom went on to great fame. They had decided, *a priori*, that I was the world's worst teacher and didn't need to bother listening to me. I decided to outsmart them, so I made the course, which turned out to be extremely useful for me, on variational principles in mechanics. That was a rather unpleasant introduction to pedagogy, but it was an introduction. The students all got As, of course. On the research front, Julian told three of us that, since dispersion relations were still the big thing, he had a couple of suggestions. Wally Gilbert, whom I met in Bern at the Relativity meeting two years earlier, had since earned his PhD with Salam at Cambridge. He was fated to earn a Nobel Prize, not in Physics, but in Molecular Biology. The other person in our trio was George Sudarshan who, with Bob Marshak, had been one of the originators of the V-A theory of Weak Interactions. He was extremely bright and quite sure of himself. We were roughly on a par at Physics level and talent. Indeed, we turned out a series of reasonable papers over that year, ones that are still cited.

I also had the special pleasure of finally being on the right side of Julian's lunch breaks. I was one of the happy few entitled to go out with him after his course, when he escaped from the horrors of graduate students. We would discuss all sorts of Physics issues. It was one of the great privileges of my extended postdoc year. There was also the added attraction of the Boston Wednesday seminars becoming intelligible. I also re-established contact with Dick Arnowitt, who was now at Syracuse. Dick and I decided that Quantum Electrodynamics was pretty well done and we had no major ideas regarding Weak Interaction phenomenology, let alone Strong Interactions. So since Spin 1 was finished, the next step on that ladder would be Spin 2. [Little did I know what role Spin 3/2 would have in my later life.] Of course, we knew that meant Gravity. Our work culminated in a paper, which in itself was not super path-breaking, but provided the basis to go on to bigger and better things. We both understood Schwinger's new ideas in Field Theory, so that Linearized Relativity, that is Free Spin 2, was a perfect test-bed on which to apply them, and we did. The title included a promising Roman numeral 1, indicating that this was the beginning of a series. Little did we know that it would go into double digits. John Wheeler, who by then had become interested in Gravitation, heard about our efforts, because in those days preprints were mailed out on (terrible) mimeographed copies to various labs, where they were displayed. [Modern life finally arrived with the internet and arXiv, Paul Ginsparg's gift to the world, which instantly posts almost every paper submitted. ArXiv levels the playing field in terms of accessibility, whereas formerly the mimeographed preprints would only be sent to some few institutions. Now the product is getting out of hand again. It used to be said that *Physical Review* would soon expand faster than the speed of light because the volumes were getting longer as more and more people published. That is nothing compared to the electronic submissions site, which has long passed the million mark.] Wheeler invited us to tell him about our work, which we did happily. Once again, I traveled to Princeton, this time neither as a sightseer nor as a postdoc. We trooped into Wheeler's office. He had set up a large tape recorder and one of his trademark bound notebooks. After a few minutes, he said that given what we were beginning to tell him, he thought it might be of interest both to us and to his very gifted young graduate student, to come and listen to our presentation. We didn't protest, and in marched Charlie Misner, whom I had briefly met at Chapel Hill earlier that year. In the course of the presentation, it became clear that Charlie was pretty smart and knew a lot that we didn't. We didn't really know much about Relativity. We knew a lot about Spin 2. At the end of the session, Wheeler suggested that it might be of mutual benefit for us to collaborate on extending the Hamiltonian analysis we had suggested to the full theory. Indeed, Charlie had shown us one or two of his results, which seemed to be of particular importance. We airily said we'd let him know, since at this stage in our glorious careers, a mere graduate student was not really of much interest. On the train back we discussed it and realized that it would be utter stupidity not to accept this offer; Charlie was happily admitted to the fold. That's when the acronym ADM replaced plain AD. Our work extended over a period of three years, at which time our interests and geography diverged. We remained on very good terms thereafter. Dick died not so long ago. Charlie is still going strong, now Emeritus at the University of Maryland. Our results provided an alternate, modern, version of General Relativity that was to become a touchstone during the theory's subsequent explosive expansion.

It was at this point that I realized how Chekhov's dictum—that he had to squeeze the serf out of his blood drop by drop—applied as well to me: I had to squeeze the student out slowly before I could aspire to original research. Indeed, this was a turning point. I was finally able to join the ranks, generate original ideas and work them out at the level to which I had aspired so long. This happened in my mid-20s, though Martians like Julian (inhumanly brilliant) were fully formed at sixteen. One's most original contributions are not necessarily made in one's earliest years. It certainly happens, and it certainly happens in Math, but in Physics one has to absorb a little more. If you look at the fathers of Quantum Mechanics, or Niels Bohr's discovery of the atom, they were all at least in their mid-20s, which was roughly when I came back to the US; Schrödinger, the oldest of the bunch, contributed considerably later than that. At the other end, we have examples, particularly in Mathematics, like Euler and Gauss, who were productive into very old age, especially considering the state of medicine in that era. The message is that it is possible to start early, possible to start late, possible to have dry runs, and also possible to have a brilliant first successful paper and then fizzle. Clearly, nature samples just about every possibility and there are no guarantees!

BOSTON—ADULTHOOD

Brandeis University was still setting itself up in 1958, having been founded a decade earlier. Science is expensive. The Physics department, because there was little money or space yet, was almost entirely driven by the Theory division. There were minimal labs and certainly no experimental research. The faculty consisted of four people, who went to Julian for his recommendation for the fifth position. On his say-so, I was appointed Visiting Associate Professor with a three-year contract, and expectation of tenure if all went well. We decided not to commute from Cambridge but to buy a house in Waltham, Brandeis's location. We saw one that seemed interesting, on a little peninsula on the Charles River, which flows below Brandeis through Waltham on its way to the ocean, past nearly every institution of higher learning in the Boston area. Knowing we would be away for the summer, we enlisted a local lawyer to buy it for us.

There was to be a small Gravity workshop in Neuchâtel, Switzerland, underwritten by the US Air Force research arm, then at Wright Field in Ohio. They sponsored a great deal of gravitational research and did an enormous amount for its establishment in the US. Just about everyone working in the field had a grant from the USAF, as did I. In those days, USAF-sponsored scientists had to travel on Military Air Transport Service planes. My wife and daughter were to travel a couple of days after me on a regular commercial flight.

On landing in Paris, I was surprised to see an Air Force Colonel waiting for me. He explained that a French inventor was claiming that he had the secret of levitation, which I might appreciate would be of some interest to the Air Force and that I could be of great service to them in assessing this claim by watching a demonstration. Should the US not be interested, he intimated that there were other air force agencies that would be. I had just flown quite a few hours, facing backwards, the mandatory configuration on MATS, but agreed that I would show up at the Embassy the following day. In came the inventor; a Brigadier General, the Colonel and I watched as he turned on his apparatus. A mica disk took off and hovered a bit above the table. The brass were very eager to hear my opinion. I replied that the greatest service I could render to the United States was for him to be picked up by the Soviet Air Force. They were shocked by this evaluation. I explained that magnetic fields could, under the right circumstances, lift a little mass above the ground, but that it scarcely had a future on the kind of macro scale they were looking for. "Would you put that in writing for Washington?" they asked; I replied that it was my duty as a citizen to do so. I can only hope that the Soviet Union picked it up.

I had acquired a Citroën, known as a Deux Chevaux (two French horsepower). Its nickname was *citron pressé* or "squeezed lemon". *Pressé* means both squeezed and squeezed for time, as in a hurry. It had a little two-cylinder engine and essentially no power, but a roof that you could roll halfway down, like a sardine tin. It did not even have electric windshield wipers; they were operated manually by turning a knob. Likewise, there was no self-starter: one had to crank the engine to life manually each time, like in the old days. At least it hardly used gas. I picked up my wife and child at Orly, and we proceeded to the meeting, held in a hotel on top of a steep hill; it was barely possible to get the car up its slope. Pauli visited from Zurich to see what was new at this dawn of Gravity's renaissance. Alas, the playwright Friedrich Dürrenmatt got wind of us, too—he lived in the city—and came by to meet Pauli, I suppose for atmosphere, while writing "The Physicists." Unfortunately for our hopes of learning from the great man, Dürrenmatt invited him to inspect his cellar: we never saw Pauli again!

My wife and I had decided, for a long-delayed vacation, to drive to the northern tip of Norway, North Cape, at Latitude 71 North (4+degrees North of the Arctic circle). It was the last continental point in the West where one could experience 24 hour days. The Arctic Circle has at least one day of 24 hours; the North Pole receives 24 hour days for 6 months. In those days the *Deux Chevaux* was the car to have because the route through Norway and back down through Finland was unpaved, except for the big cities, and gas stations were few and far between. God help you if you skipped the chance to refuel. The Citroën had no gauges. To find out how much gas you had, you used a dipstick in the gas tank. If the dipstick was dry, but you could still smell the fumes, you had a few more miles left. We reached Nordkapp, where, at 2 a.m. we watched the sun as it stayed entirely above the horizon, its lower end just skirting the ocean. A Lambretta scooter pulled up, on which were a French couple, both teachers, who admired our car as there was a two-year wait list for *Deux Chevaux* in France. We fixed a date to sell it to them back in Paris. Then we made the mistake of using our map, which had been correct thus far, since there was only one road in Norway. It showed one short-cut road back down from Norway through Finland to Helsinki. It omitted to mention that the first one hundred kilometers of this road did not exist. It was just sharp gravel. We had one hundred scary kilometers because the number of spare tires we carried was not very large. Miraculously, we made it to the Finnish border, but it wasn't until many miles later that there was a border post for the real road. These few words do no justice to our camping, near-death encounters with mosquitos, and the general wonder that we were still alive after each étape! When we arrived in Helsinki, we collected our mail at American Express. There awaited a letter from our attorney in Waltham stating that unfortunately the house purchase had fallen through, but he had taken the liberty of buying another house for us instead. No pictures, just that simple statement. Eventually we made our way back to Paris, where I put my family on their plane to the US. I had to take MATS back and transfer the car. I arrived at the designated café at the appointed time, like Phileas Fogg in Jules Verne's *Around the World in Eighty Days*. The French couple from Nordkapp were there with their whole family to celebrate. I then returned to the US to face the beginnings of my academic career and the house of our lawyer's dreams.

ACADEMIA: (UN) STEADY STATE

I embarked on my three-year Visiting Associate Professorship at Brandeis, at which point they were free to fire me. Indeed, early in my last year, I was informed that they would not retain me because I had not collaborated with anyone. No one else had collaborated either, each of us being in different fields. In the meantime, I had achieved my first major conquest in Physics—that ADM formulation, a highly collaborative project. Arnowitt was confident he could get me a job at Syracuse, not my promised land. On the other hand, it was the only game in town at this point. Until it wasn't; I was informed that Syracuse would not make me an offer because one of the Brandeis four, who had been recruited from Syracuse and was a bit of a loose cannon, simply told them not to hire me. I protested that this was ruining my career without cause. The Brandeis four finally revoked the termination. I was made a tenured associate professor. Obviously, the atmosphere remained strained for some time after this episode, but since my research was engaging, and my work more and more successful, I put it all behind me. This painful episode is recounted here for the benefit of starting physicists, though they are unlikely to face such pathology in academia.

Adjusting to the life of a faculty member at a rather atypical university was a transition, by no means all bad. One learns to lecture and to endure endless meetings. With time, one is able to endure fewer of them. Science was not yet central to Brandeis. Graduate students, on the other hand, began to flock there precisely because they did not fit into the normal mold. Indeed, we discovered in a number of cases where this applied fairly dramatically. One graduate student took to burning crosses on campus at night. Others got caught up in the 1960s student uprisings. There were people who had dedicated their lives to esoteric parts of deep philosophical Quantum Mechanics. We also had some very good ones. I was happy to acquire a great number of absolutely first class postdocs. Having good postdocs and graduate students keeps one on one's toes: one has to have ideas to propose to them. We were part of the greater Boston community; the Harvard, MIT, Tufts, and Boston University Theory departments hosted rotating lectures and seminars. Boston has always regarded itself as the Athens of America, if not as they say there, the Hub of the Universe. Aside from certain climatic drawbacks, such as endless winters, it is a wonderful old-line area, surrounded by spectacular countryside. It was the wrong place for agriculture, but certainly the right one for really charming surroundings and conservation land. There are lots of parks, beaches and walks. We raised our family in these ideal surroundings for close to 50 years, from 1957 until 2005. This period was one of great progress in our field, from finishing up Quantum Electrodynamics to the rise and essential completion of the Standard Model, our last great breakthrough. Other deep ideas were developed, especially the work of Ken Wilson: it explained what physical theory was not, and that what we understood at any stage was only good up to some given energy, but had washed out all the presumably deeper Physics that occurs at still higher energies/lower lengths.

During the years 1958–1961, I was almost exclusively involved in the reformulation of Einstein's General Relativity. The three of us (ADM) had the best run of luck that one could possibly hope for. Results just kept pouring out in all directions. Most of them were not immediately appreciated because they were of the kind that would be understood by Quantum Field Theory and High Energy people, who were not yet interested in General Relativity, while traditional Relativists had no interest in Quantum Field Theory. Our work's significance took some time to emerge, although a few concepts did catch on quickly, like the so-called ADM Energy. Suffice it to say that we achieved a complete reformulation of Einstein's original geometrical General Relativity into a "normal" field theory.

My next international conference was at Royaumont, a converted abbey near Paris, in 1959. There, Dirac and we (ADM) each expounded our then new, and to some extent similar, results. We also did at its successor, in Jablonna, Poland in 1962 whose organizer was Leopold Infeld, a strange, brooding figure; he was Jewish, but lucky enough to get a Rockefeller grant in the late 1930s from Poland to work with Einstein, which saved his life. He taught in the US and Canada, but fell for an offer to rebuild Polish Physics under the Communist regime; in part, he fled the McCarthy era's horrors. As we landed in Warsaw, several late model Western cars were lined up. They belonged to the Polish Relativists who had gone abroad and were allowed to use their earnings to bring them back to Poland. Everyone was vying to drive Feynman to the Polish Versailles. We went into Warsaw once or twice. I was shocked that what had been the Ghetto—truly sacred ground—had been essentially erased. Instead, there were playgrounds and apartment buildings, with nary a reminder of the Holocaust.

There is a coda to this story of early travel. Soon after, I was visited by a crackpot sort of person, trying to sell some insane Gravity variant by a Finnish mathematician. I gently explained why it was wrong. He was involved with the US Army Transportation Corps, because a few days later, there appeared a Captain in my Brandeis office, who told me they were grateful for my input. He inquired if there was anything they could do for me. I said, "I already have a grant from the Air Force." Then I thought about CERN, which was just beginning to get going in Geneva. I said I would not mind spending a year at CERN, but they would have to pay my salary and expenses. "No problem," he said. "Just send me the budget and justification." So that encounter gave me a productive and interesting year at CERN.

INNOCENTS ABROAD

CERN, which had finally moved into its new site, was still a work in progress, especially in terms of its experimental facilities. The Theory division was already quite large and very good, since Theory is cheap. They had recruited primarily from the CERN signatories, mostly in Western Europe, but with visitors from everywhere. Geneva is a beautiful city, although its weather leaves something to be desired, with a spectacular countryside: on our side the Jura, a relatively low altitude range, made it that much easier to enjoy. Across the lake were the high Alps. Everybody was completely bitten by skiing, so on any given weekend in winter, there was almost no one left in town. I found the atmosphere very productive. Faculty children, because of the travel on sabbaticals, are used to going overseas, usually to a pleasant destination. During that year away, children get to attend school using another language. Ours learned French on a variety of occasions, and it has stuck. They got to see how other schools work, not without a certain amount of frustration. The quality of the housing was variable and highly dependent on the host institute. We had to make no allowances in Geneva that year, however. We lived in a suburb, Satigny, on "the Farm", owned by one of the big Geneva families, with a charming yard overlooking the Alps. One could take the train into Geneva or drive. I could walk to CERN through a gate in the backyard. The Farm had all the amenities. The village still had its own cows, whose milk we would collect daily. Once there was a big Swiss scandal: the cows around Zermatt had caused a typhus epidemic. I asked the milkman in our village "Is there any danger here?" He recoiled and said, "No, Monsieur. We haven't had a case of typhus for at least five years."

Less picturesque, but a demonstration of travel's broadening effects, was the beginning of a long-lasting, highly productive, partnership with David Boulware. Although I had gotten to appreciate him earlier, in

Boston while he worked under Julian, we never collaborated until the more exotic *aegis* of CERN. We continued to work frequently and productively over the decades after his move to the University of Washington.

My next extended leave was a real sabbatical in 1966–1967, this time in Paris. We lived in Bures-sur-Yvette, a suburb on a commuter line about a half hour from Paris, in housing provided for visitors by the Institut des Hautes Études Scientifiques, although my appointment was at the Sorbonne. IHES was the French equivalent of the Institute in Princeton, serving primarily mathematicians, with somewhat fewer, but excellent, theoretical physicists—at that time Louis Michel and David Ruelle. One neighbor was Alexander Grothendieck, a legendary and mysterious figure, who ended up off the grid in the Pyrenees; one of his last mathematical works was called Children's Drawings (don't ask!). Another, Pierre Deligne, became a pied piper to all the visitors' children because of his angelic character. In later years, he was made a baron of his native Belgium, which required him to design a shield; his included three chickens, embodying the tautological nature of Mathematics. An old French nursery song says "when three chickens go afield, the first leads the way, followed by the second, while the third goes last." I was partly on a Fulbright Fellowship while there. Invitations would come from other countries that had Fulbright programs. I visited Greece on one, where after lecturing, my wife and I took a couple of days on the Greek Islands and got caught on Rhodes by the Colonels' 1967 coup. The visit was meant to be short, so we left our children in the care of our au pairs in Paris. Our separation only added to the anxiety caused by the chaos and violence of the takeover. After a dramatic interval, foreigners were allowed to leave, which we did, rapidly.

My "sole" duty was to run the de Broglie Seminar, which took place at the Institut Henri Poincaré, the IHP. "All you have to do is give a talk a week." I imagined that I had to orate every week about what wonderful research I had done the week before; that is no joke. I was aghast, but if that was what was done, then I, too, should do it. It wasn't until nearly the end of my visit that I understood my mistake; all the organizer was required to do was to round up locals to speak. I was young, and ended up giving a series of weekly talks on my current research, an exhausting endeavor. Furthermore, it turned out that the main purpose of the seminar was to bring people working in the field together once a week, not its content, as nobody had real offices in Paris. They all wanted the chance to talk to each other, "*discuter*", before, after and sometimes during, the seminar. A crowning event of this foreign adventure occurred when I thought it would be a good idea to publish something in a French journal. A colleague pointed out this would be my historic chance to meet Duc Louis de Broglie, the senior statesman of French Physics. He was the Perpetual Secretary of the French Académie des Sciences. He fully deserved this lofty position as one of the fathers of Quantum Mechanics: in 1923, he proposed what is essentially wave-particle duality. His older brother had been an eminent experimental physicist and the family was one of Europe's most aristocratic. Every week for the past four hundred years or so, the Académie publishes its journal, the *Comptes Rendus de l'Académie des Sciences de Paris*. "*Comptes Rendus*" means Proceedings. It met on Mondays, in its magnificent building on the Seine in the middle of the city. I arrived one afternoon and found de Broglie's office. The door was open and my eyes fell on de Broglie in full Academicians' garb, a green uniform almost completely encrusted in gold. He was standing at his desk and said, "I'll be with you in a minute. I just have to finish this preface to a book." He was known for being able to crank out things like that non-stop, standing up, without a single erasure, as I saw with my own eyes. Then he turned to me as I explained who I was and that I ran his seminar and had been told to meet him to submit a recent piece of work that might be considered for the *Comptes Rendus*. He looked at me suspiciously. Although my French was faultless, it was clear that I was a foreigner. It was also clear that he had never heard of me. I assume he did believe I was the de Broglie Visiting Professor. He inquired, "Has anybody else read it?" By that he meant any of his incompetent students, who had grabbed the choice positions at the IHP. I admitted that a couple of them had and they were the ones who suggested that I approach him. He became much friendlier and said, "In that case, I will be glad to have it printed in *Comptes Rendus*, but I must apologize in advance because, as you know we come out every week, but I might not be able to squeeze it in the next issue. It might take two." To one inured to the then rules of the *Physical Review*, where six months was a blink of an eye, this was amazing. However, I knew that the right answer was that I would be willing to bear the wait, if absolutely necessary. Sure enough, the following Monday my paper was printed, a truly impressive turnaround. He then buckled up his sword, adjusted his two-cornered Napoleonic hat and politely escorted me to the main auditorium where academy members were assembled. The Académie des Sciences, like ours in the United States, covers all disciplines. The proceedings began with de Broglie introducing visitors, who all bore gifts. For example, a representative of some exotic academy brought an egg of an extinct species, another visitor offered a fragment of ore. Then came the painful part of being exposed to various members lecturing on whatever they wanted to have included in the *Comptes Rendus*. The only event comparable to this occurred much later, at a conference in the Paris Observatory, the center of the astronomical world in the 17th, and to some extent the 18th Century. We met in a room where the transit of Venus was captured in a mural on the ceiling by one of the great masters. (I actually had the pleasure of seeing the real transit of Venus a few years ago in Pasadena when I looked at the sun through the appropriate black glasses. You see the little dot of Venus going all the way across the face of the sun.) In Paris, Venus was represented in human form, going past a variety of quite interested male astronomers. The painted version was visually much more scenic.

I was often reminded as I lived this lofty intellectual existence, that it was hardly fated. For example, at the end of the 1966–1967 sabbatical, I was urged to lecture in Madrid, then in the last years of Franco's reign, but with a new generation on the move, thanks to CERN. I boarded a *Wagon-Lits* in Paris and after a few hours of very comfortable travel, arrived in Hendaye. I never had to even lift my head from the pillow as a porter collected my US passport (sans visa!) and returned it with a bow. What a contrast to that time, 27 years earlier, when I first crossed that border into Spain with my parents. I cannot express the emotions this transit evoked in me—and in my parents when I told them—their little boy running for his life had become a respected scientist, traveling first class, carrying the weight, not of gold coins, but of that earlier trauma.

In spring 1968, Abdus Salam and his International Centre for Theoretical Physics (ICTP) in Trieste had the brilliant idea of gathering the old guard, the survivors of the Quantum revolution, who would each talk about their past. I was lucky enough to attend. The meeting included

Dirac, Heisenberg, Klein, the whole surviving galaxy of the super-great. To us this was an amazing experience: too late for Bohr, Einstein and Fermi, but still dazzling. Perhaps the social high point was a candlelit banquet in the palace where Rilke wrote his Duino elegies; it was owned by the Thurn und Taxis, a noble family with medieval roots who held the mail monopoly in the Holy Roman Empire.

Heisenberg was, rightly, shunned by his peers. Unfortunately, he had also caused damage to post-war German Physics (whose leader he was by default), as he had lost touch with modern developments. He had gotten involved with some baroque nonsense (a bit like de Broglie), with which he even managed to infect Pauli, until the latter finally disowned it in a dramatic lecture at the 1958 Physical Society meeting at Columbia, an occasion that few attendees will forget. Worse was his blackballing of the new generation, particularly LSZ: Harry Lehmann, Kurt Symanzik and Wolfhart Zimmermann, who were to be so important in the Quantum Field Theory renaissance. Mrs. Heisenberg was, if anything, even more tainted than her husband; many years after the war she published her memoirs, whose tone can be understood from the fact that she had the *chutzpah* to complain that in the last days of the war, her children were sometimes deprived of fresh milk for a day. Surprisingly, the book had a preface by Viki Weisskopf, himself a Jewish refugee from Vienna, and a world-class theoretician, who was later plucked from MIT to direct CERN. We were good friends, so I complained about this *lapsus*; his only excuse was the characteristically Viennese reply that he could hardly refuse her when she asked him.

The Trieste lectures were an inspiration to us attendees and should be required reading for students, who are not otherwise exposed to historical development. It would greatly benefit them to learn some of the insights these people were sharing, looking back on their youth. Once in a rare while, we still hear someone from older days, but depredations in our older generation have only left a very few such people.

From Trieste, I felt obligated to honor an invitation to Prague whose all-too-short Spring had just begun. Those were truly heady times, as I realized when I was waved through Passport Control on arrival. The physicists were as elated as the general population and the continuous festivities were in proportion to the decades of oppression. A few days later, just after I left for Vienna, the tanks rolled in; I saw the stream of

Czech refugees there, while attending the international high energy conference, overseen by an aptly named Miss Steel from CERN.

Given the rules on sabbaticals, I was eligible for another in 1971–1972. By this time all the Paris physicists had gone to Orsay, the stop before Bures on the commuter line. There simply was no more space in Paris. We again lived in IHES housing, although my visit was officially to Orsay, known as the Université Paris-Sud. I would have an even shorter walk, just a couple of blocks, to my office. That was a high-powered community. The French had finally shaken off the burden of de Broglie and his crowd and were turning out excellent original scientists. Two who were most relevant to me were Eugene Cremmer and Joel Scherk. The former never failed to give me essential insights and suggestions. Both were seminal in early String research and went on to become bulwarks of Supergravity; Joel was very mild-mannered; he once wrote me that he was "rather furious" that some French ministry had messed up his travel grant to lecture at Brandeis at my invitation. There is no such thing as a free lunch, so they asked me to give a course on modern General Relativity. This suited me, except that I ended up writing it all down as lecture notes, more the size of a small book. That got quite a bit of interest and helped usher the subject in as part of the modern curriculum.

During term break, January 1971, I accepted an invitation from Carlos Aragone to lecture under Fulbright auspices at the National University in Montevideo, to be attended by Latin American Relativists. I went via Rio where there was a Conference during one of Brazil's bad periods—they oscillated between Brazil and Argentina, making the scientists oscillate as well. Little did I (or indeed, anyone) know what a fraught time that would be, the beginning of dictatorial darkness, much of it with the CIA's collusion, as I already saw when I met some of the "cultural attachés" at our embassy. Uruguay had until then been a shining example, the Switzerland of South America, with the endearing feature of an economically dictated embargo on new cars; its rolling stock made Cuba's look positively futuristic. That idyll was destroyed by the dictatorship that followed and forced Aragone into exile in Venezuela, then an island of peace. After my lectures, I embarked on a whirlwind tour of the continent, through Buenos Aires, Santiago, La Paz, Cuzco, Lima, Bogota, Panama, and ... New Orleans. In Buenos Aires I met, clandestinely, two of their best theorists,

Giambiagi and Bollini, then lying low. La Paz and its airport have the highest elevation I had ever reached; I quickly understood why there were numerous oxygen tanks on site. Indeed, my hotel was hosting the Brazilian soccer team, the powerhouse of the world, all lying motionless on the lobby sofas! I knew the head of Bolivia's cosmic ray station from graduate student days. This resulted in a Jeep odyssey to the station, at 16,500 feet, and a virulent case of altitude sickness. I got from La Paz to Peru on the Altiplano (so named because it is above 14,000 feet), involving a boat ride across Lake Titicaca plus train, to the old Inca capital, Cuzco and its then sparsely visited neighbor, Machu Picchu. Cuzco's airport is one of the more dangerous, requiring the plane to do a sharp right-angle turn between mountains just after takeoff, all this at high altitude. Breathing Lima's sea level oxygen was compensation. Bogota, another high altitude capital, had lively university people in an active atmosphere, not to mention its unique collection of pre-Colombian treasures and gems. The next stop was the Panama canal, duly visited and admired. The trip ended in New Orleans, where I was jolted back to the US South by the taxi driver taking me to the city for a couple of hours' layover. Above his windshield was a stocked gun rack, which, he explained graphically, was to discourage any undesirables who had the temerity of hailing his cab. Still, to see the sweep of the Mississippi was to me the city's high point. It was nevertheless a pleasure to touch Boston's snowy soil at last! I revisited Brazil much more recently, with my wife and our eldest daughter, and the exotic city of Ouro Preto (meaning Black Gold, not oil), whose state, Minas Gerais, was the origin of enormous 17th Century mining riches, converted in part to Baroque cathedrals with enormous organs. We heard a (Bach of course) concert in one of them—a magical, if disorienting experience. The city had been home to the great poet Elizabeth Bishop, whose house was in its pristine state, unlike some of the old mines we visited! A few days in Rio revealed much that I had not seen back in 1971, while luckily avoiding the now rampant street crime.

A BIG YEAR

In 1975–1976, it was time for another leave from Brandeis. My wife and I decided not to uproot our children from their schools so the first semester I stayed local but split the second one between Kings College, London and CERN. The family would join me as soon as school was over, so the girls could spend the summer in and around Geneva. This turned out to be an *annus mirabilis* for my research, one in which many things came together in unexpected ways. That first semester I was appointed Loeb Lecturer at Harvard on Paul Martin's nomination. This was prestigious and fun. The Loeb is an endowed set of lectures in its Physics department, intended to explain frontier results in one's field; in my case, the state of the art in Gravity. I was also invited by Goldberger to Princeton as a short-term visiting Professor, where I worked (separately) with two then-lowly instructors, Claudio Teitelboim and Frank Wilczek. The collaboration with Claudio continued over the years, including a simple proof of the positive energy of both Supergravity (SUGRA for short) and General Relativity shortly after the former's birth—see below.

Around New Year's Day, I flew to King's College, one of the original bastions of modern General Relativity. I stayed in a "council flat", a proletarian apartment run by one of the boroughs of the City London in one of the only remaining bombed-out areas not yet reconstructed. It had been pretty proletarian all along. One had to walk through a full-scale Hollywood wartime movie set to get to it from the Tube station and climb a steep staircase, often adorned with drug users. Being alone, I worked a great deal and often came home quite late. The only place I encountered on the way was a Fish and Chips shop, where the food was always wrapped in yesterday's newspaper. I didn't eat there too often, but then again food was not (yet) the reason one came to London. At King's were a couple of very bright young people, Mike Duff and Chris Isham, whom

I'd met before. We immediately got into the swing with our research. This was the beginning of something called the study of anomalies, in particular conformal anomalies, which are a popular and important subject of Quantum Field Theory to this day. The research dealt with the distorting effects of quantizing a system that is classically perfectly acceptable. The Physics building was an example of the worst kind of London post-war construction, located just off the Waterloo Bridge on the banks of the Thames. My office was on the fourth floor. The Strand widens there; in the middle, like an island, was the ancient church of St. Mary's-in-the-Strand. The view from my room was of its steeple. What made it less pleasant was the notorious London winter chill that easily permeates flimsy post-war buildings. At some point it became so cold, even for my relatively young age, that I couldn't work. Finally, one of the custodians came by, wearing his standard grey smock and carrying a whirligig-like instrument. He held a handle, at right angles to which was a little object that rotated. I inquired if he was sent to warm up the molecules by his motion. He replied indignantly that the device was for measuring the average temperature. After a few moments of its agitation, he took a reading, and said, "Sir, the temperature is 40°F (5°C)." I inquired if that was "a little bit below normal." He said, "Yes. We will try to do what we can." In due course, the temperature went up to 50°F (10°C). I never really warmed up until I left in the spring. At some point, I was giving a seminar and there appeared Hermann Bondi, about to become Sir Hermann. He was Her Majesty's Chief Scientific Advisor to the Department of Energy. He arrived late with one of those battered government briefcases chained to his wrist. I was told that was a rare sighting. He was a smart, stimulating man who led some of the early advances in the field.

It was with regret that I watched April 1st approaching, since we three were going great guns. On the other hand, it was not exactly torture to be moving on to Geneva's springtime. I went via Paris to recover my old Citroën, which we had bought several years before and left in Europe. After a long day's drive, on pre-autoroute roads, I arrived at the CERN cafeteria around 9 p.m. The experimentalists worked all night so they had to have a place to get food and stay awake. The theoreticians often worked late, so they showed up as well. There was Bruno Zumino, one of my old friends and colleagues. Indeed, I had invited him to give a set of lectures at the last of the Brandeis summer schools, in 1970. These lectures turned out to be extremely prophetic. I was in touch with Bruno a great deal. He and Julius Wess, his close collaborator, who was also at NYU, discovered something called Supersymmetry (SUSY) around 1973. Actually, they had independently re-discovered it, since Gol'fand and Likhtman in the Soviet Union first did in 1970, but no one in the West paid attention. No one in the West paid attention to Wess and Zumino either. No one, including myself, understood what point it might have. This is something that is now a main object of searches at the Large Hadron Collider. Indeed, it was one of the reasons the LHC was commissioned: an interesting example of how Physics functions. At the cafeteria, Bruno said he wanted to talk to me about something, though I was pretty exhausted. The next three hours we sat and discussed the possibilities of Supergravity. I had also thought a bit about it, but for entirely different reasons. He had been awaiting my arrival because I would be the one person at CERN who would be able to collaborate. We spent 18 hours a day at the blackboards. He was eight years older than I, so that was an even greater task for him. Our best work was done in the major CERN conference auditorium, a huge room whose advantage to us was the infinity of blackboards where we could write our extremely complicated formulas and work them out. In a period of just three weeks, to our amazement, we had a consistent theory. It hinged partly on the fact that both he and I had learned things in our past that nobody else knew. We were thereby able to short-circuit enormous calculations. There was competition: three other people were working very hard on this and were just about finishing up; it was neck-and-neck.

I had arranged to go home and visit my family on April 23rd. At the time, some one-plane airline was offering flights from Geneva to Luxembourg, where the new Icelandic Airlines had its European headquarters. There was barely room for us two passengers. The pilot and co-pilot navigated using Michelin road maps, trying to find landmarks below. Terrible thunderstorms meant we couldn't fly very high. Both of us were in mortal terror until we finally touched down in Luxembourg. The rest of the trip was, by comparison, an Air Force One experience. Bruno and I had just finished our (quite short and entirely analytic) manuscript as I left; it was typed up by the incomparable Tatiana Fabergé, chief secretary of our division. Bruno handed it to Raoul Gatto, Editor of *Physics Letters*, while the other trio also finished their lengthy, and computer-assisted, calculations. There were some rather unpleasant priority aspects to all this, which to this day have left a nasty aura. The other team claimed that we had merely improved on their results, rather than independently achieved them, and employed various non-Physics means to try to convince people of this manifestly absurd and self-serving claim. I freely admit it rankled us both for quite a long time, especially since it gained credence among people who should have known better, simply by reading the papers! Bruno and I wrote other works on Supergravity; one was the first formulation of Superstrings.

CERN expands enormously over the summer when people from all over the world come and make it a very pleasant milieu. The population of the Theory division used to swell even more. In the office next to mine was a very bright, hardworking applied mathematician from Harvard, T. T. Wu, who told me that he and Yang had a problem that was driving them crazy. This was the question of proving that there could be no static solution of Yang–Mills Theory, just as there could be no static solution of Maxwell's Theory. It is often said that you learn subjects in Physics by teaching them, certainly true in this case. I remembered having taught students in my mechanics course that a static system cannot have self-stress. I realized, as I sat in some seminar, that this 19th Century theorem provided a one-line proof for the Yang-Mills system. The theorem had sort of been proven by Coleman and by Derrick around the same time, much more ponderously.

Back in Boston, I had two very bright graduate students, Kellogg Stelle, now professor at Imperial College, and John Kay. Kelly was finishing his thesis. He came to Brandeis in an unusual way. He was an undergraduate at Harvard, and wrote us asking for an early decision to be admitted as a graduate student as he was about to go off for a year to the South Pole. He would stay until the ice melted and since there was no internet, he would not be getting mail, if at all, on any reliable basis. We happily took him on. John was also getting close to finishing. We started a variety of Supergravity projects. In particular, one of the reasons for betting on Supergravity was that it might allay, if not entirely remove, the infinities that made the direct quantization of General Relativity inconsistent. Stelle and Kay were very bright, so they picked it up and got excited about the project. There are successive steps to it, and Pure Gravity passes only the first one. The next was a different story. We discovered that the next step was also safe because there were no possible two-loop infinities in Supergravity, unlike, as had been very laboriously shown, in pure General Relativity. So Supergravity was an enormous improvement from Pure Gravity: it was finite in the first two orders. We were ready to send off a paper to *Physical Review Letters*, the prime rapid publication journal, saying Supergravity is finite. We relied on Fermi's "Theorem": if its first and second approximations work, then a claim must be true. This is of course a rather weak "theorem", so I decided we had better wait till morning and sleep on it, literally. This illustrates the sometimes byzantine paths/forks involved in seeking new results. During my sleepless night, I recalled that matter (SUSY), *not* SUGRA, systems have supermultiplets led by the system's stress tensor, hence invariants starting with its square. Gravity has no stress tensor, however; its nearest avatar is the Bel-Robinson tensor, quadratic in curvature and only covariantly conserved. I guessed that it must have a SUGRA continuation, in which case the above SUSY picture is reproduced, but with a resulting scalar ~(Bel–Robinson)^2, namely, quartic in curvature—just the feared three-loop death knell. Those nightmares had parallels, Stelle tells me, in his and Kay's feverishly seeking quartic terms in the then-existing General Relativity textbooks that same evening. It is now known as "the notorious DKS curvature^4 term": our brilliant delay saved us from ignominy. Those were days of great fervor. As new American Institute of Physics keyword classifications, *Supergravity* and *Supersymmetry*, acquired a niche of their own.

Neither Supergravity nor Supersymmetry show any sign of being used by nature; with every passing year that seems less likely: It does not have to use anything we invent. We all like to say that elegance is so important because in its economy and clarity it leads us towards the light. Dirac always believed that. Sometimes, however, we only recognize a theory's elegance after it has been proven to be correct. On the other hand, another great hero of times gone by, Ludwig Boltzmann, one of the fathers of Statistical Mechanics amongst other things, used to say *"Eleganz ist für Schneider;"*—elegance is for tailors. He didn't care how much heavy lifting you had to do if it got you the story right. Ours were very elegant theories and they have kept a lot of people busy. It must be remembered that all of this research in the mid-1970s was in the shadow of the final developments of the Standard Model of Strong and Weak Interactions, including quarks, asymptotic freedom, the Weinberg Electroweak Theory and a number of others. This was also the glory time of CERN as they discovered the weak neutral currents, the most convincing proof of the Standard Model, at a detector aptly named Gargamelle, the mother of Rabelais's Gargantua. Super's future is hard to predict, despite the by now more than 15,000 papers on SUGRA alone. To use Chou-en-Lai's legendary reply when asked about the effects of the French revolution, "it's too early to tell."

OXFORD *ET AL.*

My next British experience, courtesy of my old friend Dennis Sciama, was an invitation to visit All Souls, the quintessential Oxford college: it does not admit students, only postdocs and up. Dennis and Dick Dalitz were the two token scientists in an otherwise humanities and law-oriented institution. Science was a black sheep, reluctantly included. There were numerous illustrious Fellows; I was, unsurprisingly, struck by (Sir) Isaiah Berlin, still active in that spring of 1977. I only had time to stay one term. All Souls is an extremely well-endowed institution, with wine cellars that would put any three-star restaurant to shame. As their (full-time) wine steward explained, their connections in Germany date back to the 1400s, only interrupted by a couple of unfortunate lapses. There was just a High Table since there were no low table denizens. This splendor contrasted with Dennis' Astrophysics group, which had a lot of smart people, all encased in a by-then permanent temporary Quonset hut. One felt as though one were inside a submarine. I was used to that from my graduate student days, but it was still depressing. I even watched the obligatory cricket match, although I must admit I was not fully caught up with its glamour. John Sparrow, the Warden, was an old-line literary person, about as eccentric as one could imagine. When he drove his ancient Jaguar, all the natives fled for cover. I knew this was going to be an interesting term when the porters, the people who do all the work at the colleges, explained that not only was I permitted to walk on the lawn across the quad at All Souls, but that I was required to do so (shades of *1984*). Robes were *de rigueur* at meals; fortunately (a moth-eaten) one always hung in a certain spot for us visitors. At some point they decided to introduce a college tie. After a long debate as to the design, they chose Mallard ducks on a blue background, for the legendary giant Mallard that flew out of the foundations when it was being built in 1437. Indeed, there were statues of these ducks, inset in the magnificent library holding six centuries of books. Upon presentation of credentials, we could purchase a tie. The shopkeeper inquired if I was a Fellow of All Souls. He accepted my affirmation, though indicating he knew I was lying. There were two models, polyester and silk. "I'm the last of the big American spenders, so I'll go for the silk." But was I a Permanent Fellow? "No, I'm a Visiting." "Then I'm only permitted to sell you the polyester." When dinner was over, one would take Port. That ritual consisted of moving to a smaller room where, on the round table, a carriage bearing the decanter was wheeled around the table.

Special occasions at All Souls were called Gaudies, which many ex-Fellows attended. These feasts began with liberal quantities of champagne. Some of the jurists were retired, others worked at high levels of the British legal system. Isaiah knew a little of my history, that as a kid I had spent a few months in Palestine around 1935. As we were talking, along came one of those boilerplate retired overseas civil servants. As Isaiah explained, this gentleman had been the Chief Justice of the Palestine Supreme Court under the British Protectorate in the 1930s. "The two of you will probably enjoy sharing your experiences." The fact that there was about a fifty-year gap in our ages didn't seem to bother him. The judge asked where I had been posted. I had to disillusion him. Still, it was fun having gone from a scabby kid in 1935 Palestine to a colleague of its Chief Justice.

The wives of some of the Fellows had pressured the College to allow women to attend a couple of dinners each year. As it happened, my mother-in-law arrived from Sweden during one of those times. She was not eager to attend one of these misogynist occasions, but I convinced her that the tourist value would make it worthwhile. It was not the Gaudies I knew; we were served a terrible meal, with not very good wines. The main aim was apparently to deter any future visits by females. We were outraged. Times have changed: Women are now Wardens, as well as a large fraction of the Fellows. Another example of modernization is illustrated by how the other Fellows then regarded Physics. A university professor, famous in his field, once asked me to take a little turn in the garden. He blushed and confessed that he had come for advice. He didn't want to confide in the locals. His teenage son was showing signs of wanting to become a physicist. That was why he blushed. I had the feeling that he wanted to know if this strange proclivity could perhaps be deprogrammed. I asked a few questions and assured him it would not bring shame on the family name if indeed things went in that direction. I'm not sure what finally transpired, but he seemed greatly relieved.

I did have one other noteworthy incident at Oxford. Dennis Sciama had asked me to give a couple of lectures on Supergravity. The other Oxford theory group was led by Roger Penrose, a truly original force in Mathematical Physics. His boys were intensely loyal. Roger could not attend, but his people filled the front row. At the end of the talk one of them asked, "Has Penrose seen this and does he agree with it? Also, what has this got to do with Twistors?" Twistors were one of Roger's elegant mathematical discoveries. To the first question, I could only say, "You have to ask Roger." To the second question, it turned out much later that Twistors have indeed been used in connection with Supergravity and many other things. Another example of their mathematical bent occurred a bit earlier at Kings College when Roger's boys believed they were work-ing on real Physics. One of them gave us a couple of talks on his latest. At the end of the lecture, he concluded with: "So we have proven that there can only be two charged leptons: the electron and the muon." I shyly raised my hand and said, "Actually, it has recently been established that there is a third lepton, the Tau, a more symmetric idea." "That's impossi-ble. What is it?" "You can look it up. It is a very massive companion to the other two. It should be there, corresponding to the triplets we see in other places." He muttered and went off. A week later his talk ended with a bombshell: "So we have shown that there are exactly three charged lep-tons," to which I burst into wild applause.

That year also included a Summer School in Cargese. I knew nothing about Corsica, except that it was a marvelous island. I was co-director, with Maurice Lévy, one of the post-war French theoreticians. Corsica is never a quiet place; there is always some sort of rebellion, mostly regard-ing independence from France, and always a lot of shooting. During the Nazi occupation, the word Maquis described those who served in the underground; it's actually a word for the hinterlands, which are scrubland. In Corsica, too, the rebels occupied the maquis. The cities were mostly on the coast and were controlled by the French, but by no means entirely.

I flew from London, destined for Ajaccio, on the southwest side near Sardinia, which it almost touches there. Instead, the flight ended up at the antipodal point, Bastia on the northeast side, closer to Tuscany. I was the only passenger for Ajaccio, so Air France provided me with a car and driver to cross the island to Cargese. The driver enjoyed explaining to me where it was dangerous and where we were likely to meet the rebels and what they were likely to do to us, so it was a fairly agitated ride. We never saw any rebels, which seemed to disappoint him. My family joined me soon after; we stayed in an apartment above the local bar, which was hotly contested by the two sides as evidenced by the many bullet holes. The housing in Corsica was something of a come-down after Oxford, but certainly not in terms of climate. The beaches were fun, and the Physics equally so. The third of the European Summer School circuit was in Erice, an ancient Sicilian town perched 800 meters up on a rock to fend off aggressors. Sicily had been invaded by just about everybody, especially Normans and Saracens. It was a charming place, restored to its original splendor, with the added benefit of plumbing. The school was run by a Sicilian physicist, Nino Zichichi. As things typically go in Italy, everything worked because Zichichi was a friend of the Pope, of the reigning party, and tolerated by the Mafia, which held the countryside. There was a coast road linking Erice to Palermo, but this was disputed territory. There was a tacit agreement between the authorities and the Mafia allowing the road to be open to traffic by day, but at night it was Mafia turf. At the end, I had a very early flight out of Palermo to get to Rome and on to Tel Aviv. This required leaving Erice at some ungodly hour, in Mafia time. Nino told us not to worry; it was all arranged. I was going with Ken Wilson. He lived in an old Volkswagen camper, a derelict-looking vehicle which would not tempt anybody. We made it to Palermo airport and through security which was essentially nil. The agent saw I was heading to Tel Aviv; he had been told to check the baggage especially well, and then afterwards seal it with some kind of wire mesh. This would ensure the luggage would get through Rome without any further inspection. "I assume there is nothing in your suitcase." "Don't you want to check?" He said, "Why? You don't look dangerous." The Israelis detect terrorists trying to board planes by subjecting all passengers to a psychological interview. In the busy summer season they would hire college undergraduates, trained to ask questions. On my trip home, the student asked what I did. When I told him, he said, "What is Laplace's equation?" I was a bit supercilious. "I know, but do you?" He said, "Not really, but we were told it was a clever question to ask." So I gave him a five second seminar on Laplace's equation; his only aim became to get me off his beat.

THE EIGHTIES

The '80s were busy and highly successful. They started with a CERN sabbatical, when I worked with my Brandeis colleague Howard Schnitzer's stellar students, Paul Townsend, now a Cambridge professor, and Larry Abbott, now a Biology professor at Columbia, as well as with Warren Siegel, who had briefly been a postdoc with me. When home, I worked especially with Roman Jackiw at MIT and with Gerard 't Hooft, when he was Loeb lecturer at Harvard. That led to my most-cited papers, a bit too esoteric to detail here, involving the Chern–Simons terms, a deep mathematical construct due to the eminent mathematician SS Chern and Jim Simons, now better known as the preeminent "quant" on Wall Street. His foundation is better at supporting the basic sciences than the NSF. The work with Gerard and Roman was quite different, finding fruitful Physics in a seemingly humble area of Geometry/Gravity. 't Hooft, being a Martian, would come up with elegant solutions to our problems, where we were content with more pedestrian ones.

The Institute for Theoretical Physics that Walter Kohn had founded at Santa Barbara was beginning to get its name on the map by 1984. Bruno, who had by then moved from CERN to Berkeley, and I, were asked to run a semester's research cycle there. We chose Quantum Field Theories and General Relativity. Santa Barbara is "California's California," the ideal place in terms of weather, wealth, and ease. It certainly had those characteristics. We rented a UCSB faculty member's house in Hope Ranch, "Santa Barbara's Santa Barbara." Hope Ranch had its own private beach access, although we were never fans of the Pacific Coast as much as of the Atlantic, Mediterranean or Baltic. The house was surrounded by avocado trees. We had to take care of two (broken-down) dogs and a (ditto) car. We didn't realize that avocado groves were sources of crawling creatures that made life difficult for the dogs, whom we had to ferry to the local vet with some frequency. A number of my younger friends were there, including David Boulware, Kelly Stelle, Philip Candelas (Roger's Oxford successor), Mike Duff, Chris Pope, as well as the late Peter Freund from Chicago. It was a lively time for both Physics and our shared group dynamics.

In May 1986, a Nobel symposium on the progress in String Theory was held on the tiny island of Marstrand, a few meters off the Swedish coast. Gell-Mann was going to give the summary talk, and came around to ask us for details in an uncharacteristically humble way. I also had a chance to hear his uncanny command of languages, as we walked round the island while he recited one of H. C. Andersen's tales in flawless Danish, a language I speak well enough to appreciate, though he had no ties to Denmark. My French may have been better than his—I think he wanted to impress me because I knew the ur-meaning of the French word "*métier*". He succeeded.

THE IRON CURTAIN I

I must include one essential chapter, given the century I lived in, about the horrors of the Nazi and Stalinist crimes perpetrated on their scientists. China has its own notorious history, but there I know much less, except the outlines of Mao's persecutions of intellectuals. By now there are many books on the subject, both by historians and by victims, but I will just add a few notes. The Nazi story is summarized by the following: When Einstein was asked by a colleague going to Germany after the war if he wanted to send greetings to any colleagues, he said "Yes, to von Laue." "Anyone else?" "von Laue," he repeated. Max Planck, who died in 1947, had a son executed after the plot on Hitler and was one of the few decent scientists who stayed. I was fortunate enough to meet Lise Meitner in her later years when she had retired to Cambridge, England with her nephew Robert Frisch, with whom she had worked out the basis of uranium fission during a pre-war Christmas vacation in Sweden. She held an Austrian passport until the Anschluss invalidated it in the fall of 1938, but although Jewish, she continued her work in Berlin. She refused to see the signs and do anything about them until almost too late; it required a perilous escape to Holland, then to Sweden, where she initially stayed with the Kleins. Max Born had been forced, being Jewish, to flee to England. When Pascual Jordan asked him for a post-war "Denazification" letter, Born simply sent him a list of family members who had perished under the Nazis. The rest of the Germans were (apart from Heisenberg, a sad case) contemptible lackeys or convinced anti-Semites. A tragi-comic story concerns one of the true greats of the Quantum era, Jordan, who was impelled by his wife—she even looked like Lady Macbeth—to join the Nazi party early so as to secure a Chair, not very successfully, as it turned out. Then he joined Adenauer's party after the war for the same reason and with similar results. He might otherwise have received the Nobel Prize(s) he deserved. When I met him, he headed a separate Institute in Hamburg set up at Pauli's urging, so as not to infect the younger generation. Felix Pirani told me about the time he had agreed to talk there, after which he went to dinner chez Jordan, met at the door by Mrs. Jordan, who told him disapprovingly that his suit was not dark enough. The person next to him, seeing he was British, mentioned his dislike of Churchill. Felix, being quite left-wing, agreed, but asked why in particular? "Because Churchill invented concentration camps!" Then there was Fritz Houtermans, who had a particularly horrible life. He got it both ways, first from Stalin and then the Gestapo. He was a very good scientist, whom I met briefly: you could see the effects of his past on his face. George Gamow is another case of a great scientist who luckily got away and ended up in the US before the Russian purges went full scale. There's a story he once told regarding those years, about three people who had been very promising. He meant himself, Landau, and a man called Ivanenko. Landau was one of our greatest scientists, had his year in the cells of the Lubyanka and was nearly executed. He had recently been the victim of a car accident. Ivanenko was part of every delegation from the Soviet Union that I ever saw at international meetings. He always proposed toasts to the friendship of our peoples. Gamow said about those three people, "One is now a drunk, one a vegetable, and one a KGB slave." Indeed, Gamow was known for his heavy drinking. Looking at Ivanenko, you could see he had been broken by something. He seemed just to function mechanically, according to some script that had been tortured into him.

I was told an amusing story by the late mathematical physicist Ludvig Faddeev, my old friend and sometime collaborator. He was someone of my generation, who received great international renown. He was allowed out of the USSR quite frequently and was elected to membership in all sorts of prestigious academies. At the end of the bad old days in the late 1980s, many scientists would go on vacation to the absolutely most unreachable areas in Siberia where no public transportation existed. They would just live there picking mushrooms, hiking and climbing mountains. Faddeev was advocating at that time to turn the Steklov institute in Leningrad into an international center for mathematical Physics to be named after Euler. [Russia always named institutes in honor of people.] This idea had to go up to the highest levels for approval. He didn't hear anything and was on vacation during the most remote summer he had ever spent. He told me that one sometimes found (this is almost impossible to believe) small colonies of very old people who were not aware of the Soviet revolution. As he was hiking one day, a Soviet Air Force helicopter descended to tell him he was to receive a phone call later that day and to transport him to the nearest telephone, quite a long distance away. He was airlifted from the middle of nowhere to the phone; he knew immediately from the phone number that the call was from the Kremlin. He was told that the institute had finally been approved. He was to return to Leningrad and set the machinery going. Sadly, nothing ever came of it, for the usual irrelevant political reasons that often derail such projects also in the West—here a feud between the apparatchiks in Leningrad and those in Moscow.

Much more uplifting, was to meet the great heroes. The most famous one was Andrei Sakharov, a great physicist and a great man. He had been a believer in the Soviet system, but finally realized it was the wrong system, which immediately demoted him from the highest Lenin medal tier to exile in Gorky. I made only two trips to the Soviet Union, as it happened, two years in a row. I never wanted to go to Russia during those years but I was convinced by colleagues in Boston, the first time in 1987, as Perestroika took root. There were so many *refuseniks*: people who lived in very strained circumstances because they had applied for emigration to Israel; the Jewish ones, and having been denied permission to leave, were also denied permission to do anything but the most menial jobs with continual KGB harassment. I could bring them whatever calculators were available at that time and improve their economic status a little bit. So I agreed to go to a meeting in Moscow.

I took the subway out to a distant suburb where the *refuseniks* lived and had their meetings. It was very touching to see how these people were keeping up their scientific credentials as best they could. There was one particular couple, a really hard case. They had been fired from Moscow State University, the most prestigious institution in the country, run by Anatoly Logunov, who knew how to roll with the political punches. Indeed he was the *de facto* czar of Theoretical Physics in the Soviet Union. His permission would be the deciding factor in anything that was done for this couple. Logunov was a terrible physicist, who made his later career denying General Relativity—the old Stalinist line.

He was doing real nonsense, which, instead of General Relativity he called some idiotic thing like "Relativity General." I was going to talk on what I was doing, as is customary, but my Plan B was to debunk Logunov's theory in front of this distinguished audience. Just before my talk, he arrived with two heavies from the KGB, who sat down on either side of me and asked me what my topic would be. Apparently the rumors had gone around about what I was going to say. I replied, "You will hear it very soon. There is no point in saying it twice." They spoke in heavily accented English. Murray Gell-Mann said one could always tell who the KGB agents were at an international meeting, not only because of the way they dressed, but if asked what they worked on they were all taught to say "MagnetoKhydrodynamics." I gave one of the more inspired talks of my life, which was easy, given the nature of the material. The applause was deafening, except from my two neighbors. As I came down from the podium, the great theoretical astrophysicist and a father of the Soviet Weapons project alongside Sakharov, Yakov Zeldovitch, gave me a bear hug and said, "Moladyetz" which means "great guy", quite a tribute. Then came Logunov's turn to speak; a rather painful, if short, experience. The chairman of that session was Gerard 't Hooft. Next, one of Logunov's boys proceeded to give the details of the big picture that Logunov had just painted which went on and on; neither Gerard, nor even Logunov, could stem the flow. He just kept going: Slide 155, slide 156, ... Finally 't Hooft got up and said "I'm hereby resigning as Chairman because it is impossible to run this session in any orderly way." Finally, the robot was pulled off the podium.

Sakharov was also in attendance at the meeting. One could immediately see how revered he was. Everyone was happy that he had finally been allowed back from Gorky. The conference site had a very limited bathroom capacity. Long lines would form at breaks. Each time, Sakharov was ushered to the top of that line, a real tribute. I introduced myself, to which Sakharov said graciously, "Of course I know of you." When Larry Abbott and I were in CERN we wrote a pretty good Cosmology article, the sort of territory that Sakharov was interested in. At CERN, and elsewhere, there was a big movement to free Sakharov. We felt we should do something, so we dedicated our paper to him on his 60th birthday, figuring that this was inoffensive enough. *Nuclear Physics*

B immediately accepted it, but the editor, Hector Rubinstein, had decided one should not rock the boat *vis-à-vis* the Soviet Union. Otherwise, the Soviets wouldn't allow people to publish there. He asked us to remove it. I knew him well and was quite angry. He said, "Look. This is my position. Take it or leave it." So we took out the dedication and when the proofs came, reinserted it and Rubinstein never noticed, but Sakharov did because friends from the Lebedev Physical Institute used to visit him with the latest journals. He said, "I want to tell you that that was very cheering." Wheeler and I were invited to his apartment to meet his wife. The address was fairly central, and for Moscow it was a pretty nice place. Khriplovich, one of Sakharov's staunch defenders and helpers, was also there. He advised that both Sakharovs were away, but would be coming back soon. Then there was a phone call, and Khriplovich said he, too, would have to go out for a while. He said, "Don't worry; make yourself at home and just answer the phone." I said, "I don't know any Russian, except for about five words." He said, "Just tell them in Russian 'He is not there.'" That was within the limits of my linguistic abilities. Of course, the phone rang absolutely non-stop and I got to pronounce those words fairly well. The first person to come back was his wife, Yelena Bonner. She was an absolute fireball. This was still the Russia of all shortages. Apparently, if anybody learned something was for sale, they would alert their friends. In this case, the rumor was that a shipment of Bulgarian refrigerators had arrived and she had rushed off to get one, because theirs was on the blink. Unfortunately, it turned out that they were sold out, the usual story. She instructed me to make myself at home and she would make some chai. Sakharov duly appeared. He looked very tired, but happy. We spoke for quite a while about Physics; Wheeler arrived shortly thereafter. At that point, Sakharov had an interesting, if not quite correct, theory about effective actions in General Relativity. It was a true privilege to have met and interacted with him, however briefly.

I had told the Logunov-dependent *refuseniks* to meet me at the station the next evening, as I was headed to Leningrad, so I could tell them about what, if anything, I had accomplished.

At the end of the day's talks, I took Logunov aside and said I'd like to speak to him about something completely different. He said, "Of course, a pleasure." He hadn't gotten this far without being politic. I explained their terrible story. He countered, "You know I absolutely cannot do anything for them. I am just a lowly academic. Yes, I am involved with Moscow State University, but I don't have much authority. Of course, I will inquire and see what I can do." The next day, he came back to me and said, "I have asked, but I've been told that their case is so egregious that they may have to wait at least seven years before getting out." He was clearly lying to illustrate his powerlessness. By the tracks of the Leningrad train, I told these poor Russians that I had spoken to Logunov and that he claimed he had no influence; I didn't want to mention the seven years. I could only wish them luck—which fortunately came fast under Gorbachev's auspices.

Faddeev couldn't come to the Moscow meeting; instead he suggested I come up to Leningrad to give a talk. He said, "Getting a train ticket from Moscow to Leningrad the same day is almost impossible, but I think we can pull some strings." The next thing I knew, one of the secretaries at the meeting had a ticket for me in a sleeper car. After I had discharged my unpleasant task, I got on board. Nobody was in my cabin, but a tunic hung there. I believed, and then checked, that it was that of a Colonel in the KGB. I think Colonel was their lowest grade. After a while, the Colonel arrived and realized that I was not a fellow traveler, as it were, so he kept a cold, but correct, distance. In the morning, we were served chai from the wagon's samovar. Ludvig met me at the station. We had our usual very pleasant discussions. I had a wonderful time in Leningrad in my one day. He also took me home to his apartment house in the suburbs, at the end of the subway line. The subways in both Moscow and Leningrad would make a Westerner sick with envy. They are palatial installations, with wonderful features like two sets of doors, so that our horrible stampedes cannot occur. I also noted that people on the subway were all reading what looked like serious books.

We did get to tour individually around Moscow. We were invited to a wonderfully moving concert in one of the older cathedrals. A Russian physicist's wife was involved in preserving it. The space was one gigantic repository of icons of the highest quality. There we stood listening to archaic religious singing. I also remember a reception hosted by one of the good Russian scientists in a beautiful, old-line apartment house in an older part of the city, filled with books. The Russian intelligentsia were remarkable people. Gerard and I left late, after the tramways had stopped running. We'd been told we could just hail any car. We raised our hands and a little car came to a screeching halt. We got in and told him where we wanted to go. He wanted one dollar from each of us, which we were quite willing to part with. So that was our introduction to the underground economy.

THE IRON CURTAIN II

The following summer there was to be a commemoration in Leningrad of Alexander Friedmann, the great cosmologist who had died of typhus in 1925, soon after his fundamental Physics contributions. My wife and I were going to be in Sweden at the time of the meeting, so Leningrad did not seem far away. This was 1988 and the situation had already improved a little with *Perestroika*. A visa was to be arranged for us, together with tickets to get to Leningrad from Helsinki. We had a car and ferried over to Helsinki. The next morning we went to the Russian Embassy, which as one can imagine, was one of the most imposing—if ugliest—buildings in the city. We walked in and were told to speak to the Scientific Attaché, who was drunk at 10 a.m. We explained that we were there to pick up our visas and tickets. He had never heard of us; "What tickets?" I told him, "We're going to the meeting sponsored by the Soviet Academy of Sciences." He said, "Too late. No tickets on plane. No tickets on train." So I said, "Well, I guess that sort of covers it. We'll just go home and explain to the Academy that you couldn't get us there." He said, "No. No. Don't worry. Of course, I can fix." He took our passports and led us down some secret underground corridor from Section A to Section B. He inquired, "Could you get there by car?" I replied that yes, we did have a car, but were not permitted to use it in the USSR. He said, "Don't worry. Give me your license plate number. I'll just put it in here in your passport." I replied that I could do that as well, but that didn't seem very official, besides, we don't have a visa. He shooed us away, "Don't worry. I will call ahead and take care of it." There was only one border crossing from Finland for cars. We were supposed to be at the hotel later that day. This was right near Midsummer's Day, the holiest of all holy holidays in northern Europe. We drove on clear, beautiful, Finnish roads. As we got to the Finnish side, the guard said, "You do know that the World Cup is going on and the Russians are doing badly? So lots of luck!" We said, "You may see us again very soon as we don't even have visas." So he said, "It is very strange, but you never know." So he opened the gate and out we went. On the Russian side was a fortified installation with three different lift gates, all of which seemed to operate manually. A poor recruit came out and opened all three, so we could pull up to the customs post. We were greeted by a Captain. We explained that someone should have called on our behalf from the Embassy. "Nobody called." So I said, "What do you mean? We are supposed to get to a meeting of the Academy in Leningrad today. The Scientific Attaché sent us. You can either call him or if you don't want to, we can just turn around and go back." He shrugged, "No. No. Don't worry. We believe you." I said, "You can call the Academy." He said, "No, it is okay." We said, "We haven't eaten anything since breakfast." He replied, "You will find wonderful rest houses on the way." And suddenly there we were on Soviet territory. The first kilometer consisted of the most beautiful red carpet-like roadway I have ever seen. It was a marvel of construction. We remarked that if this was the way it was going to be, we would be in Leningrad in about half an hour. Hardly had we said that, when the road changed to a narrow, pitted surface full of treacherous potholes: we were back in the USSR! After a while we came to a big arch over the highway on which was perched a man with a gun. The light was red, so we stopped. We looked up at him, he looked down at us and beckoned for us to proceed. I was not going to; I pointed to his red light and finally he understood. Off we went. Of course there was no food whatsoever to be had. We finally stopped at a strange layover point filled with Stakhanovite Soviet mosquitos, where we ate a chocolate bar we had brought with us. We finally arrived at our modern hotel, somehow intact, if visa-less.

Our visit coincided with what the Russians call White Nights, the most celebrated evenings of the year, especially in Leningrad. All the bridges swing open over the harbor and rivers, and every battleship is lit up. There was a full display of fireworks. The spouses of the participants were given some interesting excursions, including trips to see the Summer Palace and the Peter–Paul Fortress Prison. We did not fail to explore the Hermitage despite the brutal heat. I was one of the few foreign participants at the meeting: No wonder, given what it took to get there.

A few kilometers into our trip back, a truck was parked, filling our entire lane. Just as we tried to pass the truck, a policeman stepped out and flagged us to the side. Obviously, this was a trap. We pretended not to speak Russian. He pointed out that we had crossed a double yellow line. We pointed out that there was a truck in the way. We had hardly any rubles left, but did need some to buy gas before we got to the border. He mentioned some trivial amount. In turn, I pointed to my wallet and said, "No rubles. Deutschmarks. Dollars." That scared him and he waved us on. We were stopped a couple of more times with similar set-ups. Each time, we weren't worrying about the trumped-up fines, but that someone would ask us for our (non-existent) visas. Then came the quest for a gas station. They were not too hard to identify. A woman sitting in a little tower presided over an ancient pre-war gas pump. You paid her and then she would crank by hand the equivalent liters into a glass container that you discharged into your tank. She sternly warned me that what I didn't use would not get a refund. Gas was practically free. The big problem was the octane. She explained that we were lucky because we had landed in one of the high-class gas stations. The gas was maybe 80 or 82 octane, not quite what Saab specified, but we figured we could make it with the remainder of the proper gas we already had. I then made the mistake of asking for the bathroom. She said, "This is a gas station, not a lavatory." We finally arrived at the border. There was nobody there. The gate booms were all down. We had heard that this was the deciding day for the Soviet soccer team who were playing some semi- or quarter-final against Holland and were likely to lose. The same officer we had seen before came down very grumpily, checked all our papers without too much fuss and raised the booms so he could get back to his television. They were also watching in the Finnish customs station but the mood was quite different. So was ours, being back on Western soil! We still marvel at our stupidity in undertaking this illegal, dangerous, journey.

TIANANMEN

One would not pick three years in a row for such adventures, but then an offer to go behind the Bamboo Curtain did not come frequently in those days. It was all due to TD Lee, one of our great theoreticians. He was the 1957 Nobelist for Parity Violation with Frank Yang (the fastest recognition ever given, namely the year after their paper), both members of that UChicago post-war "all-Nobel class." They were also the only two students brave enough to take Chandrasekhar's General Relativity course there at 8 a.m. in the winter quarter. Both had become revered in China in those hopeful late '80s. [Lee and Yang had a sad falling out, one that induced a heartrending Samizdat by TD called, "Broken Symmetry."] I knew him well in his Columbia years (and of course from the Institute before); as an old New Yorker, I cannot forget those dangerous '70s when the city seemed destined for oblivion. Once, after I had given the weekly seminar, we all walked up Broadway under the elevated to his favorite Chinese restaurant, the Shanghai on 125th, being assured that there was no crime there. Then a car backfired; I have never seen such a speedy retreat to the nearest doorway as by those locals!

In order to encourage the growth of Physics in China, Lee had organized a special summer school for the brightest Chinese students. He had support at the highest level, via the Chairman's private secretary, and felt so confident that he enlisted some of his friends in the West to lecture; we, in turn, could hardly say no. It was quite a list, including Sidney Coleman, Roman Jackiw, David Gross, Gerard 't Hooft, and one of the Zamolodchikov twins from the USSR. To sweeten the pot, we were even encouraged to bring our families. As the time approached, the student protests began heating up; for some reason (oldest if not wisest?) all the participants decided I should make the decision about whether or not to go. When the day came, I spoke to Lee, who had just hung up with the locals and was assured there would be no violence. I accepted this retrospectively meaningless promise, and we all packed our bags. We took our children on a new flight from San Francisco to Beijing via Shanghai, where our papers were scanned in some primitive hut on a small airfield. Beijing's was hardly bigger, so it was easy to be met, then driven to the magnificent Imperial Palace just outside the city, complete with charming guest cottages, a pleasant dining hall and a tent for our lectures. This was meant to be a three-week conference but, alas, it didn't work out that way. At the end of the first week, during which we had the chance to visit the growingly bold student encampment in Tiananmen Square, discreetly surrounded by police, we were suddenly taken on an impromptu field trip to Datong. We might have been alarmed at this sudden change in schedule given that it was most certainly not on the menu. We were taken on an overnight ride in an ancient East German sleeping car to that dusty industrial city on the edge of the desert, whose only attraction were some enormous Buddhas (indeed impressive) carved into the surrounding cliffs. That (Saturday) night, we were loaded back onto the train for Beijing, again with no explanation; by then we were suspicious, since Datong had clearly been totally unprepared for us. Even had we not been worried, already in the far suburbs, we sensed something was wrong, with smoke rising from many spots inside the city. At Beijing Central, there was no longer any doubt. Needless to say, no bus was awaiting us; TD, ever resourceful (and perhaps remorseful) managed to hire us an unmarked van that was willing to risk the trip but was too small to take all of us at once. We fearfully voted on which triage choice was safest and decided to have the dependents go first, lest things got worse. Nothing was moving in the city; cars, buses and streetcars were overturned, some of them burning or riddled with bullet holes. To our infinite relief, the van returned for us after a long time, telling us our kids had made it back safely. The trauma of waiting so long for any news did not fade very quickly! As we drove on the highways back to the Palace, we passed scenes of heavy armed action and troops with rifles pointed towards us on all the overpasses, but we still had no idea of what had occurred. As we sat around our campus waiting to see how we would get out of the country, students who had been at the Square began arriving with news of what had happened. They urged us to bring their stories to the West as they were sure that the authorities would suppress the truth. It was of course the end of TD's school, if not (yet) of us, for of course nothing flew into or out of China. I should have said almost nothing: Yuval Ne'eman, who had shown up from Israel, even though the two countries had no official relations, had somehow been spirited back to Hong Kong, we never knew how. Later, a wise Belgian physicist, Leon van Hove, who had earlier run CERN, explained why I should never have trusted the Chinese assurances. We were later told that our students managed to return to their home bases safely. It was an all-too-thrilling bit of involuntary war-tourism, even including a sumptuous Peking Duck banquet welcome at the start. TD heroically managed to get us all out on the first flight of the partially reopened airport, which was a total madhouse, to Tokyo, the nearest stop. On arrival, we were greeted by TV cameras and flocks of reporters, images of ourselves we later saw in our hotel. We were much too shaky to go on to the US directly, and opted for a couple of days in the city. That was not so easy, as there seemed to be no rooms that night. I was given a pile of Yen coins and a phone directory. We landed a business hotel, a euphemism for minimal area, but clean and adequate. We did whirlwind tourism, unencumbered by guide books, but managed to see Frank Lloyd Wright's old Imperial hotel before it was demolished, as well as the major sights, eating great food in obscure local eateries totally devoid of English, and even a subway rush hour ride (luckily before dinner). So ended my last adventure of its kind.

A LAST SABBATICAL

The year 1993–1994 was split between the Institute and CERN. That first semester was darkened by the demise of the Superconducting Super Collider, the last big US accelerator project, derailed at least as much by Senate politics as by cost overruns, just when hope for high energy progress was running high. We left for Europe on New Year's day, and moved into a CERN apartment in Grand Saconnex, just across the street from Geneva airport's runway—extremely convenient for us, being linked by a pedestrian tunnel. That spring, my wife and I simultaneously caught the worst flu since Chapel Hill (for me) which immobilized us for weeks. On the other hand, I was awarded two honors during that season—election to the National Academy of Sciences, and the prestigious Dannie Heineman prize in Mathematical Physics, for ADM, which is presented at the annual APS meeting back in the US; luckily, Arnowitt and Misner could be there to accept. I felt especially proud because the head of the selection committee was Nambu. The National Academy ceremony was almost a year later, so I was able to attend and sign in. When our CERN visit came to an end, we undertook a *Mitteleuropean* voyage, having finally overcome our Geneva illnesses. First stop was the Swiss Ticino: in Ascona, near Locarno, lies a conference center from the 1920s, set atop a hill aptly called Monte Verità, overlooking Lago Maggiore, the largest of the Alpine lakes. It was run with full Swiss precision, meals timed to the minute. This sojourn gave me a chance to expand on earlier work on Conformal Anomalies, a subject then unfamiliar to Relativists. At the end of the session, I was made conscious of my advanced age by being appointed to give the traditional closing thanks, this from the balcony of an excessively picturesque rustic inn, making me feel like a (bad) Operetta tenor. The next stop was Vienna, where I was to serve as Schrödinger Professor at his new, eponymous, Institute.

94

We rented a small, but central, apartment across from the Opera, near the Sacher (of Torte fame) Hotel, and as we discovered, right next to W. H. Auden's old *pied-à-terre* on the Street of the Whale. Lest it all feel too Gemütlich, the city's many bookstores were all featuring Kurt Waldheim's autobiography. To be fair, a modest monument to the perished Jews had recently been installed in the old Jewish square. Austria's Nazi role in the Anschluss is well summarized by the old remark that only Austria could convince the world that Beethoven was Austrian and Hitler German. Its anti-Semitism is an old tradition, already present in the old dual-monarchy pre-WWI days, but in less murderous form, as that Empire had many nationalities to manage, a world masterfully depicted in Robert Musil's *The Man without Qualities*, with—to be sure—parallels in much of Western Europe. We even attended Wagner's *Götterdämmerung* at the Opera, in a suffocating heat wave—those poor singers were dressed for the eternal *Dämmerung*. To see Boltzmann's— and all the great composers'—monuments in the gigantic Central Cemetery and the early 20th Century architecture, art and design masterworks of the *Wiener Werkstätte* and the Secession, was a cultural education. After that month and its lectures, it was time to drive, via Prague, Dresden and Potsdam, to Sweden. After passing Austerlitz—the crowning battle of French children's Napoleonic lessons (I did see Waterloo at some point, to even up), we passed the Czech border, where a very different picture emerged: a narrow, crowded highway full of trucks, sad women displaying their charms at roadside, and a parade of East German, Romanian and Czech cars. We passed Theresienstadt, a so-called "mild" extermination camp; that impressed itself on our consciousness. In Prague, the office of my host in the Academy of Sciences was in the same building and very room in which Kafka had toiled at the National Insurance Bureau; it gave us material understanding of his darkness. We could appreciate the city's many other charms as it was not yet overrun by mass tourism. Dresden had been almost entirely rebuilt, including its Royal art gallery. Potsdam was chosen because it houses the Albert Einstein Institute, an extremely active center, as can be imagined. Its director, Hermann Nicolai, was another old CERN collaborator. We also admired Erich Mendelsohn's (surprisingly small) Einstein observatory, built to measure his famous red-shift.

We enjoyed several shorter stays in Paris during those years, one in connection with a winter school at IHP—my old haunt—but this time I had a bona fide office and key, and a lively audience. It was also a lesson on the city's hidden topography: from our apartment house, shoehorned into a cul-de-sac in the aristocratic 7th Arrondissement, our view encompassed an erstwhile 16th Century Nunnery, complete with a garden of that era. Beyond its wall stretched Michelin's headquarters. I also re-learned a Paris geometry theorem: between any two points, there runs a bus (+/– a few blocks).

Lest it be supposed that a physicist's life is one continuous travelogue, I emphasize that I am eliding all the heavy lifting: the time spent teaching and its preparations, seeing students, Faculty meetings, emptying overflowing wastebaskets of failed calculations, refereeing, writing letters of recommendation, enduring boring lectures, endless grant applications and evaluating proposals of all sorts from all over the world. So it is not just, in Coleman's paraphrase of Thorstein Veblen, "The leisure of the theory classes." Indeed, this account should be taken in the spirit of the old Israeli joke: A saintly man goes to heaven, where he spends eons lolling on clouds, but eventually wonders what he might have missed down below. His guardian angel agrees to the visit. So down he goes, where he is amazed by the endless orgies of all sorts indulged in by happy beings. Back in heaven, he cannot help but miss that action while on his boring cloud. Finally, he asks his angel for another visit. The angel informs him that only one is allowed, after which it's a one-way trip. Some eons later, he finally decides to take the plunge. As he falls into a lake of boiling oil, stabbed with pitchforks, he wonders what happened: "Last time you were a tourist, now you're an immigrant." This account is the tourist version!

MEETINGS AND MEMORIALS

Perhaps surprisingly, attending many of the shorter memorial celebrations and meetings where one's lectures are a sometimes expensive price of admission, is a taxing activity. As an extreme illustration, I recall the case of Andrei Linde, who was suddenly chosen by the KGB—in the bad old USSR days—to attend a conference in Florence. To any Russian, that would be the impossible dream come true. However, Linde felt he had nothing significant to say, and told the KGB he would not go. This comical impasse of the KGB forcing him to go to heaven and his defiance of that order was not taken lightly; fortunately, he came up with Cosmic Inflation just in time (as did Alan Guth, independently) and all went well. That may have been the same meeting when the city's Communist Mayor spoke to us about Galileo and the Church, on its very steps! Italy is indeed a paradise, where I regret never having spent extended time; where else would a great masterpiece, Piero's *Madonna del Parto*, be visible in a country churchyard, using a key from the nearby vestry (I hope it's safer now)? I have attended Solvay conferences in Brussels (and once in Austin), with their historic aura from the old pictures of Einstein and his colleagues, though unfortunately we no longer dress photogenically. A number of other meetings merit special mention, such as Oskar Klein's 1994 Centennial in Stockholm, where his manifold ideas were at last recognized, alas too late for him. Instead of being in charge of the Nobel Committee, he should have been a Laureate; indeed he was often nominated for the many achievements named after him: the Klein–Nishina effect, the Klein paradox, Kaluza–Klein theory and Klein–Gordon equation among others, not to mention discovering the Schrödinger equation just as he was felled by jaundice and unable to publish before Schrödinger did! Apparently Schrödinger began to work on this problem after a lecture on de Broglie's recent work on waves. As they were leaving the room,

Debye shouted to Schrödinger that, as the expert on wave equations, he should figure out the one governing de Broglie's. The rest is history—Schrödinger retired to a chalet (with his girlfriend!) and did so! I have been lucky in my choice both of parents and of parents-in-law: Oskar and Gerda were most kind, warm, and cultivated, and their company and support were never-failing; their other children, likewise, a pleasure.

2018 saw Feynman's and Schwinger's Centennials. Both were prodigies of the New York public school system. Feynman once entertained me, on the New London ferry to the Shelter island II conference (the original, back in the late 1940s, marked the first breakthroughs of QED, including theirs), with his transition from Far Rockaway High to MIT, where he and a fellow-student (Weldon) taught themselves Quantum Mechanics; he seemed very proud of that. Such prodigies often leave collateral damage for their immediate successors: Feynman had rhapsodized about Mr. Bader, who inspired him in high school. I later met a very smart physicist (Martin Annis), also from Far Rockaway, who founded one of the first private high-tech enterprises, and asked him if Mr. Bader was also his idol. "Don't mention that so-and-so; he nearly ruined my life. I needed his recommendation to get into MIT, and he told me I was nowhere near as good as the last graduate he sent there!" I believe something like this also happened to people who were in school with Dyson. Ironically, Feynman really wanted to go to Columbia, but the Jewish quota had been filled for his year.

There were several celebrations for Julian, both when he was alive and after his all-too-early demise in 1994. As might be imagined, these commemorations were bittersweet, recalling the past and the inevitability of our heroes' endings. Still, it forced me to consider Julian's influence and accomplishments as they impinged on my own work, let alone the bigger world. One such memory involved the genesis of Supergravity and my being sent for by Julian to tutor him over a weekend when it was just new; an apt closing of a circle.

There was a small, but very touching, gathering for Melba Phillips at Brooklyn College, when she was still alert at 90; we alumni reminisced about what she had taught us in preparation for real Physics. After all these decades, the campus seemed both small and dangerous, with metal detectors at its entrance; many of us trekked there that weekend morning.

I went with another alumnus, Aaron Galonsky, one of our high school era gang, who went on to an esteemed career in Nuclear Physics. The other, sadder, kind of meetings are also to be noted—where the honorees were dying—first for Sidney Coleman at Harvard in 2005, then Bruno Zumino's 90th in Berkeley, then Gell-Mann's quark anniversary at Caltech, all wrapped in forced good cheer.

By far the most exotic meeting I attended was held at the Chilean Antarctic research station... in deepest (South) Polar winter, in the late 1990s. It was organized by Claudio Teitelboim and used the (post-Pinochet) Chilean Air force for logistics. I flew into Santiago—an endless trek from Boston—then by military 707 to Punta Arenas, the world's southernmost city, on the Drake channel across from Tierra del Fuego in the Straits of Magellan; there we transferred to a Hercules cargo plane where we hung onto netting during the long flight to the station, landing on skis in unbroken darkness, reminiscent of the more somber engravings of Jules Verne's books. We were told that the return flight might be delayed up to two weeks by unpredictable weather. Naturally, Stephen Hawking did not miss this adventure, including some extra private helicopter trips. Such surroundings could not but eclipse the Physics, although my old Cambridge University colleague Gary Gibbons and I did manage to write a decent paper there. As if that was not enough, since the Air Force needed flight hours and we did leave the station on time, we got extra hops in one-engine planes landing on unmanned grass strips in Chilean Patagonia, a super-Alpine landscape experience of its own. The 707 back to Santiago happened to pull up next to a departing American airliner, so I simply strolled onto it—a whole day early. That trip did qualify as a Veblen.

DISTINCTIONS

One unavoidably picks up honors by living long enough, but sometimes there are shortcuts. As a second-year graduate student, in 1950, I became a member of the American Physical Society (APS). There was another category called Fellows, who paid $15 instead of $10, in dues. That narrowed the field: Who was going to pay an extra $5 for something of dubious value? In the late 1970s, the APS decided to inaugurate Divisions. Mine was the Division of Particles and Fields. We had to form the first Board, name a President, Vice-President, and other officers. I was duly informed that I had been selected to be a member of the Board; I thought it a singular honor, but quickly discovered that it was what the French call a *cadeau empoisonné*. The division's first action was to make sure that the practitioners of Particles & Fields got adequate recognition within the APS. I was charged with the tedious task of going through the whole APS membership catalog, to discover which of our Theory members was not a Fellow, but ought to be. Bill Willis, an experimental physicist of about my age and level, was to do the same on the experimental side. Sure enough, I discovered the usual egregious absences. Bill did the same. We submitted these names to our apparatchiks on the Board. They ordered us to send those lists to the Secretary of the APS, as the division's nominees. We got an icy response, saying that our nominations could not be acted upon, since neither of us was a Fellow. We forwarded this information to our handlers. About a week later, we each received a letter from the president of the APS, stating that at an ad hoc session of its Council, we were promoted to Fellows; should we wish to nominate anyone else to Fellowship, we were invited to do so. Nowadays Fellowship has become a desirable trophy, simply because it is deemed hard to get. Another honor unexpectedly came my way in 1979—election to the American Academy of Arts and Sciences, founded in the 18th century. When I first came to Boston, it was located in a townhouse on Newbury Street, now the prime elegant shopping area; indeed that house was converted into some chi-chi beauty salon. After moving to other venues, including a suburban estate, it is now permanently housed near Harvard and has a global reach. There are positive aspects of Academy membership. I was presented with a little rosette a number of years ago, to emulate the French Légion d'Honneur that people who still wear jackets sport in their lapels, which reaped two benefits. Once I went to a meeting in Paris during a time of terrorist worries. Not that I looked especially guilty, but I decided that perhaps sporting the rosette would help. At the baggage carousel, three nasty-looking police with submachine guns scowled at every incoming passenger, but then they looked closer and noticed my rosette, they waved me on with what passes for a smile in that branch. The other incident occurred when we were in a two-star restaurant. Its patrons were mostly well-heeled locals. This was the traditional Sunday lunch, attended by dowagers and literati, all of whom were greeted with a warm welcome and a handshake. It was crowded, so although we had a reservation, I was worried that we were going to be treated poorly. Instead, we got the same treatment as the regulars. As we prepared to leave, the host came up to us and asked if I would be willing to reveal to him what my rosette was, because it was not one of the several gradations of Legion d'Honneur he was familiar with. "It is nothing, Monsieur," I said, "just a mark of membership in some foreign Academy." That seemed to make him even happier because it was exotic.

Along with the Dannie Heineman Prize, the National Academy, and honorary membership in the Torino Academy, a more recent honor was the Albert Einstein Medal co-awarded to me for ADM. I could no longer undertake the journey, so I sent my grandson to Bern, then on a semester abroad, to represent me, in a borrowed dark suit. Should a similar occasion recur, I will travel via video, a more civilized mode.

With a cast and locale similar to the Nobel ceremonies, I was awarded an honorary D. Sc. at Stockholm University's Centennial at Stockholm's City Hall, including the Royal family and the King handing out the Diplomas. The obvious difference was climatic: it was held in a heat wave, in late May 1977. My crimson Harvard robes were made for arctic New England, so I decided to risk it with just underwear beneath. Next to me was (Sir) David Wilson, then head of the British Museum, who confessed to the same sin beneath his own colorful Oxford ones. There was a rehearsal the day before, as one had to walk backwards down several steps, lest one commit *lèse-majesté* by showing one's hindquarters. Henri Cartan, of the French math dynasty, did not deign to practice: Having over thirty degrees, he needed no guidance. [At All Souls, I met a professor of Canon law who beat that. Since many universities require the presence of one of his ilk at their commencements, he would be awarded multiple honorary degrees every year.] I was to be exposed to the same royalty at a subsequent honorary degree event in Gothenburg, at the price of giving a short description of our universe, in Swedish yet. I had had an earlier contact with the Queen: after giving a talk, also in Gothenburg, I was on the last plane to Stockholm. It was surrounded by police cars, a rarity in those days. She and her attendants were heading home, but being numerous, occupied the Coach section, while we proles were forced into business class. I expected attentive service, but nary a glass of water was I offered, as the stewardesses hovered over a soccer idol.

In a slightly different vein, a more dubious honor came my way some decades ago. The NSF was facing a proposal for what became LIGO, but whose initial budget was already quite high. Marcel Bardon, who headed NSF's Physics section, decided I would be a useful propagandist, given ADM's focus on gravitational radiation. I was as ardent an advocate as I felt, but did caution it would be a long haul to pay off; how long, no one (fortunately) realized. [Rai Weiss told me, when his Nobel prize was announced, that he well remembered my brief.] There was no doubt of gravitational radiation's reality, only of its detectability with a finite budget. I gave it all I had. It was to be a long gestation period indeed, with an ultimate cast of thousands (of authors) on the initial announcement. As with any breakthrough, it has now become a workaday tool all over the globe, both observationally and theoretically.

As I view the final proofs, the Royal Society informs me of my election as Foreign Member; a rare honor: there are only about 180 worldwide, in all sciences.

FESTSPIEL

A Festspiel is a classic academic celebration of a scholar's life and accomplishments. It is an old European tradition to mark retirement or a significant birthday or degree. Three of my best students, postdocs and collaborators, Mike Duff, Kelly Stelle and Richard Woodard, conspired to hold one in my honor 51 years after my PhD, in 2004. There was even a Festschrift of scientific contributions published in conjunction with the celebration (complete with my picture on the cover). Since Mike was then the first Oskar Klein Professor, it was aptly held in Ann Arbor. Richard had been passed on to me for PhD supervision by Sidney Coleman (who was not Gravity-minded), a great acquisition and an ornament to Quantum Cosmology, with whom I keep contact and collaboration as well. Kelly came from Imperial, whence Mike had been recruited. To be so honored does have its rigors: I had to sit in the front row at every lecture, stay awake (not being Dirac), and pretend to ask deep questions, for several days in a row. A Festspiel is an emotionally charged occasion, as can be imagined, because one sees the evolution of people from one's distant past. That was a fitting way to conclude my Brandeis career: we headed West to California the following year. This move also reunited me, some sixty years later, with my old Brooklyn friend, Noel, Emeritus at Caltech.

With age, if one is lucky, research can continue, sometimes even at a decent level: experience compensates for the loss of computational *sitzfleisch*! Caltech offered me a guest appointment, which meant I had all of the benefits and none of the duties of a faculty position: I could attend the seminars and discussions of interest, work on any project that struck my fancy and publish at will. Indeed, I am still surprised at the sheer volume of output this past 15 years; the quality has not been too bad either, judging by citations. Most of the work continued my tradition of long-distance collaborations on topics ranging from Quantum Gravity to Observational Cosmology to Quantum Field Theories. More concretely, I mention three: higher spin problems, uniqueness of General Relativity and partially massless fields. They are of importance to our understanding of these larger areas, and are commensurately popular and complicated. I am still attempting to do research, with results commensurate with my aging brain.

Only the Fates know how many forks in the road still lie ahead of me, but I leave this message to readers just starting out: make sure you take every one and enjoy them all. With some luck and perseverance, you may yet reach a vision of the next layer of understanding our universe; even if not, the ride will have been worth it!

APPENDICES (DEEP TRUTHS)

Niels Bohr: "A deep truth is one whose opposite is also a deep truth."

Herewith, a miscellany of more or less profound remarks about some peripheral, as well as fundamental, issues.

I. What does it all mean?

People are drawn to Physics for many reasons: it's fun, challenging, exciting but perhaps the most frequent one, I suspect—certainly mine—was "What does it all mean?" It takes considerable experience and sophistication to realize that this is an ill-posed question, a purely human reaction evolved in the context of human interactions, with no correlative in the inanimate world around us. Indeed, religion is frequently a response to it, an attempt to bring back a human-scale response—via one God or many or a whole animistic set. That this is a circular notion has not stopped its popularity through the ages. It is a typical category error, attempting to apply a concept outside its area of validity. Does this mean that Physics—as the attempt to describe and find regularities in the observable universe—is meaningless? Of course not; that would be the same error of applicability. Our interest is perfectly meaningful... to us!

The very idea of how (or if) our universe began, some 13 billion years ago, is hardly easy to swallow. I think many don't understand that the Big Bang is not some seed that exploded out of a chaotic background, but rather that there was no "there" there, no background—no "before". Likewise, the competing notion of a cyclic universe that ever repeats, possibly with altogether different laws governing each cycle, is as hard to grasp. Yet there it is, as far as we can understand right now in terms of our limited brains. The rules of this universe will look very different as we probe it on further scales, and may even apply only to some overall limited domain. Indeed, think of how far we have come from the early Greek world view, or even Newton's (a mere 400 years ago). Whether we find all this satisfactory or depressing is not an objective matter. Personally, given that we have no choice of universe, I prefer to make lemonade and marvel at this picture, however primitive it will seem to later generations. [The great Wolfgang Pauli did not make lemonade; indeed, the story is that when he went to paradise, his first stop was to explain to God how the world should have been created!] This little planet, whether or not unique for harboring some sort of "life", exists and has allowed for the evolution of biological matter called people. In a short time, they have been able to construct observational apparatus spanning countless scale generations, along with explanations and codifications of these observations from unimaginable fractions of centimeters to over ten billion light years. The present laws have emerged primarily over the past century or so: QED, Nuclear Physics and General Relativity, not to mention what is over-modestly called the Standard Model of quarks, gluons, leptons *et al.*, that governs all matter.

To summarize this compressed, technical, excursion, its message is that instead of attempting to anthropomorphize the universe—or what our instruments can see of it—via religion or other constructs, we should instead marvel at its (subjective) beauty, complexity, and our luck at the coincidence of being here to enjoy it! I must also emphasize that much of the fun and any real insight can only be obtained by becoming a pro and getting into dirty and often obscure technical calculations, as is the case with any worthwhile vocation, so any popular account is necessarily limited in what it can transmit.

Let me end this section with a Niels Bohr anecdote about coincidences. He was once taken by his students to see an episode of *The Perils of Pauline*, an early silent movie series. Afterwards, they asked how he had liked it, as a fan of Westerns. "It was OK, but I just couldn't swallow all the coincidences." "Oh, which?" "I accept that the villain tied Pauline to the tracks just as a train was approaching, and that Tom Mix happened by just in time to untie her, but that there was a cameraman on the scene to film it all was just too much!"

II. Prizes—taking the cake

The question of prizes, medals, Academy memberships, honors in general and their role in the scientific enterprise is a topic we seldom discuss with others or even think of ourselves; it is there in the background though, even if not the elephant in the room. The arena of Theoretical Physics is simultaneously extremely cooperative and competitive. The cooperative nature is manifest, since we all stand on the shoulders of our predecessors, let alone of our peers. We look at the postings of our colleagues and attend conferences to keep up with their work. This also keeps us on our toes and can be of great help with our considerations. Also, there is an economic side. As a graduate student you may only be thinking about getting your darn thesis over with, but then you begin to realize that in the process you are also involved in building your scientific and economic future. You will need to get postdoc positions and, even if you have independent means, eventually secure a job at a university or other institution in order to pursue research. It is much easier in Theoretical Physics, which doesn't require expensive equipment, but still the opportunities are limited compared to the number of people that enter the workforce every year, or at least the best of them. That competitiveness is unavoidable. There is also the matter of publication. Your publications are your calling card. Their value is certainly to the largest extent one's satisfaction in the actual work. Through our grueling attempts and stuffed waste baskets, we derive our main pleasures from the few instances when we hit some sort of jackpot— cracking the door a bit further ajar. Our favorite results quite often differ from those that are most cited. There are certain major hits that you are very happy with, but sometimes a paper that is of no immediate importance displays an elegance and originality that its author prizes. It is often said that at universities, because the stakes are so low, the infighting is intense. First of all, there is promotion and tenure as one starts up the scientific career ladder; these are the bread-and-butter issues. Then there are the external marks, such as invitations to lecture at international conferences, and the more or less prestigious publications where one is permitted to appear, based on what the referees think of your work. Gradually, a series of honors begin to loom. It starts with relatively minor ones, such as Fellowship in the APS, although even that has become a crazy field. Then, there are learned societies, like the American Academy of Arts and Sciences, followed by the premier United States one, the National Academy of Sciences. There are prizes administered by the American Institute of Physics, such as the Dannie Heineman Prize in Mathematical Physics. There is a whole series of named awards; then there are the international ones—honorary memberships and degrees, endowed lectures, perhaps culminating in the Nobel prize—though its monetary supremacy has been overtaken by very recent multi-million awards, such as the "Breakthroughs". Finally, Fulbrights and Guggenheims cover all subjects. Currently they are a bit biased against science simply because we are a bit richer than the arts and humanities. An example of absurdities in these matters: many years ago, Schwinger came up to me shyly to ask if I would recommend him for a Guggenheim, as they inflexibly need three support letters. It reminded me of Anatole France's observation that the Law, in its impartial majesty, forbids everyone, rich or poor, from sleeping under bridges. He did get it, as did I somewhat earlier, along with Fulbrights.

So, what does it all mean to us and our enterprise—good, bad, irrelevant? I am ambivalent: on the one hand, there is the obvious good—keeping Science in the mind of the public and the funding agencies, let alone the competition amongst institutions to capture and keep the winners. [Here I interject Jim Watson's story—his Dean at Harvard (where else?)— told him, the year he won the Nobel Prize for DNA, that he would not get a raise as he would not need one!] On the other, were all such distinctions abolished, we would be freed from both jealousies and feuds, as well as of the time lost in writing countless nominations, supporting documents, sitting on committees, jockeying for one's candidate etc. Furthermore, I don't need an award committee (including the Nobel) to tell me who matters in my field. In an ideal world, then—*Nyet* to honors of any kind; here on earth—a probably necessary evil, despite the many failings of the system, sometimes blatant ones, in making the correct choices and the overreliance by funding agencies and promotion committees on these externalities. But of course funding agencies are fallible, and prone to funding safer, less "controversial" (as if science is determined by voting) proposals. To be sure, it only progresses by being conservative, not succumbing to every transient "breakthrough" of overeager experimental (theoretical too!)—claim. So, like democracy (per Winston Churchill), our system is the worst possible, except when compared to the alternatives.

A related matter—essential for the very existence of science, especially as it unavoidably gets more expensive at the frontiers—has to do with its costs. We should, of course, be grateful for public funding, but it is not some unearned luxury. On the contrary, science is the seed corn for future progress, the most basic and prudent investment. We work 24 hours a day and yes, we enjoy every minute of it, only wishing we could be a little more intelligent. I think the waste, or "overhead", involved in our endeavors is truly minimal, and unavoidable, at the frontier; meanwhile we also train students in a whole spectrum of endeavors from medicine to engineering to Wall Street with the tools they will need for their work. Eugene Wigner is often quoted about how much scientists owe the world and the public for their tolerant support of our endeavors, but that is not right: we are neither exploiting it nor getting special privileges. My late friend Ken Johnson noted that if every use of Maxwell's equation required some tiny royalty, we would be rolling in gold. Physics has been the engine. Think also of steam and Thermodynamics, then later of the electronic revolution. John Bardeen, the co-discoverer of Superconductivity's basis, got an earlier Nobel Prize with his Bell Labs colleagues for transistors. Imagine how the world has made use of that! Those are just two contributions by one person, who was indeed in every way worthy of public support. Bardeen was known as a man of few words. As a professor at Illinois, he would have parties at his house. He circulated, but hardly spoke. At one party, during a pause in the conversation, he put a shiny little object on a stool in front of him. As people watched with amazement, he demonstrated the first transistor radio. Whether that (and its boombox successors) was a boon or not, I will leave unsaid. But remember that absent the transistor and its successors, there would be no information revolution, in any of its manifestations. Underlying it all is that "useless" jewel of the 20th Century, Quantum Mechanics: Quantum computing, lasers in all their uses, much of modern Chemistry, even unto Molecular Biology, Nuclear Physics—I could go on—are the fallout from those few 1925 pioneers. A tiny quantum royalty would do us royally as well! Oh, and did I mention Tim Berners-Lee's development of the WorldWideWeb while working on LHC at CERN? It merely changed all our lives, let alone multiplied the global economy, with nary a royalty to Physics!

Finally, I emphatically remind us that any society (especially a wealthy one) so crass that it does not underwrite decent support for trying to understand our universe—irrespective of possible profits or weapons—is not worth defending, as per the famous reply by R. R. Wilson when asked what a new accelerator would do for our country's defense.

III. Literature

Given that Physics is a major intellectual pursuit that has had a profound effect on the world, both practically and conceptually, one would think it would leave major traces in literature. However, I cannot see much. The only novel I am aware of that deals explicitly with modern Physics is entitled *The Unknown Quantity*, by Hermann Broch; it involved romance and graduate students in the Gottingen of the mid-1920s when it was the center of the world, where Quantum Mechanics had just been invented. Even it has precious little resonance to that. Various modern novelists have attempted, mostly unsuccessfully, to work in words like "black holes" or "uncertainty". A great Viennese novelist, Robert Musil, who died in exile in Geneva in 1942, was trained as an engineer and had claims to science; even he didn't really put it into his work. One might think that Friedrich Durrenmatt's famous play *The Physicists* fits the bill, but it is really a comedy about (atomic) espionage. There is a lot of so-called science writing in the popular press, most of which I find depressing. It is either gee-whiz, turned into a human-interest story, or worse still, the writing has a cutesy style and shows complete lack of interest in the substance of the subject. Sci-Fi, science fiction, is a different matter altogether. It is mostly Fi and relatively little Sci. I know good scientists who were snagged by Sci-Fi in their youth; in Sidney Coleman's case, it lasted a lifetime. Books like Thomas Pynchon's *Gravity's Rainbow* have certainly been affected by Newtonian Physics, but one would not gain any image of research in science from it. In the past, the Sloan Foundation encouraged eminent scientists to write autobiographies, a genre to be applauded, even if it is not literature. Alas, it seems to have succumbed to Gresham's law—the funds are now diverted to lesser pursuits. C. P. Snow wrote some awful books about science vs. humanities, best forgotten. He shed no light on what went on in Cambridge, one of the major venues for physicists from Newton to Dirac at two ends of the Lucasian Chair, not to mention Maxwell and Rutherford. Cambridge, and to a lesser extent Oxford, were the venues of some of the greatest conceptual advances ever made. Frayn's *Copenhagen* is only superficially a play about scientists in the atomic era, and also not "true history" as we now know from Bohr's accounts, found after the play's publication. I've probably missed some novels, stories or poems in which there is a more direct connection between the literary work and our field. I would love to hear of them. I think one can easily have a few fingers left over and count those that have major connections. One recent exception is the work of Alan Lightman, MIT astrophysicist turned novelist; a totally different one is the steady stream of superb layman's expositions and speculations by our greatest living theoretician, Steven Weinberg, of the state of our field, but of course that's non-fiction. By contrast, *Arrowsmith*, Sinclair Lewis's Young Adult novel, as we would call it now, is an impassioned work about 1930s biological research and romance that must have directed many teenagers towards a scientific vocation. In a different vein, popular science, a genre enjoying a worthy revival nowadays, is likely to snare some bright kids, just as their predecessors did. Many of the recent creators of prizes in our field are physicists manqué, having had, from childhood, feelings about the amazing fact that science exists at all and is able to, albeit imperfectly, describe that fraction of the universe that we can observe and extrapolate from.

IV. Other senses

One aspect of memory is the recurrence of sensory events—madeleines of the mind. For me it begins at three or four years old with summers on a Polish farm, getting milk right from the cow, eggs warm from the hen and wild strawberries from the ground, and the best ice cream ever, from local carts in summer. In Israel, it was corn-on-the-cob from vendors at the seaside. In Paris, of course, the hot chestnuts from glowing embers on cold dark winter days; in Portugal, sardines roasted on the beach right off the boat. In New York, Nathan's hot dogs on Coney Island's Boardwalk; in St. Louis, huge watermelon slices in the sweltering evening heat. Maine or Prince Edward Island lobsters boiled right from the traps, and the ineffable essence of the ocean concentrated in oysters roused from their beds in Point Reyes or anywhere else. No three-star meal was ever as vivid or memorable. That's also true of the reverse experiences, such as seemingly eternal childhood measles in a darkened room, or indeed, the safety pin episode during our escape from France. Novelists have explored these palimpsests of memory that we mostly repress—until we taste our own madeleine.

V. Bad science

There is much in the press about hanky-panky, cheating and scandals in science, primarily in Biology labs in need of funding. It is almost unthinkable for anything like this to occur in our field. Not so much because we are more honest or decent (as of course we are), but because it would be to no avail. Theoretical Physics consists primarily of long calculations, or at the other extreme, daring or far-out predictions or generalizations that don't pan out at all, none of which is fertile ground for cheating. The other issue is stealing other people's ideas. Not only is there a lot of self-policing, but also it is an automatic system. You post what you want to present. But if you post something that has already been covered or otherwise published, the system notices it and mentions that there is a resemblance to a previous paper. Anything that has been posted is automatically dated. In the past, they used to deposit sealed envelopes with some Academy of Sciences, to be opened either when they specified, or posthumously, to give priority for work they feared others would claim. I remember reading a crime novel by SS Van Dine, who was popular in my youth—it involved a Physics professor at Columbia and his graduate student. The only clue was a bloodstained shred of paper on which was written what was ominously called the Riemann–Christoffel tensor. I had no idea what that was, but it sounded very impressive. The professor killed the student in order to steal his work. I don't think such extreme attempts have yet occurred. I have been involved in situations when two sets of people independently come upon the same idea at the same time, simply because it is in the air. That's not at all surprising. Indeed, the only surprising thing is that it does not happen more often. (Of course, many of the ideas that are in the air, and therefore likely to be discovered independently, would have been better off remaining in the air.) There are minor skirmishes that can lead to bitterness between the parties, but not because of any objective importance. Simultaneity, a tricky concept in Special Relativity, is also a tricky concept here. Certainly when two sets of people come upon roughly the same idea and clearly are not in causal connection with each other, then the outside world usually credits both. There are cases when someone, even a distinguished scientist, "forgets" having seen something and simply carries it over word for word. Even if done on purpose, it is to no avail because the origin is so clear. On the whole then, we are lucky to be shielded from the various sins that could have beset us.

VI. Pure versus applied

This is a pernicious example of an ill-posed problem or, more prosaically, the goose with the golden eggs. When modern science began, the direct effects of scientific discovery really took off, partly from chemical innovations, partly medicinal, and then a whole spectrum of engineering applications, based on Newton's Laws. In the 19th Century in particular, Physics finally began to pay off. There is a famous, if probably apocryphal, chestnut about William Gladstone, the then Chancellor of the Exchequer, a man with little interest in science, asking Michael Faraday what use electricity might be. He wryly replied, "you will tax it someday"; *Si non e vero, e bene trovato!* Immense contributions derived from pure science, everything from assisted locomotion, the steam engine and the use of steam as an industrial tool; to electricity; communications, starting with the telegraph and then evolving into our current phase; Condensed Matter Physics and the semiconductor revolution; Superconductivity and aero/astronautics—the list is endless. Because of the (less wonderful) bomb developments, post-war science was initially well-subsidized by the government. There was not going to be much boosting from industry, except during the golden age of Bell Labs, whose costs were passed on to the consumer through the telephone monopoly. That was a wonderful institution but not really a case of benevolent industry helping science. In the meantime, Biology has become a prime site of both progress and miraculous cures, as they are always called in the headlines. It has become particularly heavily subsidized by the pharmaceutical industry, although much of the basic research is still carried on in university labs and at government sites such as The National Institutes of Health, at the taxpayer's expense. The question is how one should take into account the possible or actual fallout of fundamental Physics research on the industrial scene. Thanks to Congressional pressure, even the most esoteric NSF grants now have to pay verbal homage to their contribution' efforts for the welfare of our citizens. The goose has been laying these golden eggs with regularity, but beware killing it on some trumped-up grounds. There is an anecdote involving Stalin and the Soviet film industry. He once summoned his best film director and asked him how many movies were made in the Soviet Union in the past year. "About two hundred." "How many were masterpieces?" "About a dozen." Stalin: "Good. Next year, we'll make twelve."

Some aspects of basic science are more likely to pay off, whatever that means, in better chips, more intensive communications or more destructive arms. To judge it that way is to miss its main impact. Yes, Astronomy and Mathematics are key fields. In Mathematics there is cryptography and your credit card would not work unless prime numbers were well-understood. Astronomy is useful for GPS. The answer should be and is that that is not the point of either of those sciences. Nor is the point of the Large Hadron Collider making possible engines based on Higgs technology. [Yet, European high tech has already benefited enormously by learning to deal with superconductor production and applications by building it!] There is a continuous spectrum of applications drawn directly or indirectly from basic science. One should not frame debates on which part of this continuum is more important. Astronomy in the wider sense, including the new terrestrial observatories and satellite-based ones like Planck, has completely changed and will continue to change our fundamental view of the universe. Isn't that enough?

Science is not really expensive in any sense. Compared, say to the volume of sugar-based beverages, the obesity crisis and how much money goes down that drain, this is all relatively trivial stuff. Even CERN and LHC are a matter of some few billions of dollars, not very much in a high-trillion dollar based advanced economy, like Europe's or the United States'. It is fundamentally wrong, and can be self-defeating, to provide support for basic science on the fallout that it will generate applications. It is always the unexpected factor that makes the difference. In 1820, when Hans Christian Øersted discovered that electricity and magnetism affect each other and Faraday mapped the electromagnetic field, they were doing totally pure science. In the ensuing decades, particularly towards the end of the 19th Century, this brought about the electric revolution. Likewise, in the mid-1800s, there was interest in Thermodynamics, heat, entropy— all things that resulted from the use of the steam engine cycle. There was a really powerful evolution on both sides. The world was covered in rail-road tracks, just on the basis of steam engines, which also revolutionized ocean travel. Those, in turn, provided fuel for pure science developments, namely Thermodynamics and then Statistical Mechanics, which to this day echo in utterly different contexts like black hole Thermodynamics.

It is best not to attempt a separation, nor to make a hierarchy between applied and basic science. You could say that basic science is simply the tax that is paid by successful applications. Even that sells science short. I don't know why we humans are afraid to agree that it is worth a little bit of our world income to understand the mysteries of the universe as they unfold. It is hardly a divine *diktat* that anything purely aesthetic or relating to the comprehension of things is not enough. It is like saying that art is only good as decoration of Soviet-style buildings or as an investment for plutocrats' laundered money. There is a spectrum that goes from the purest Mathematics to quite applied engineering; they all have their place. They should not be competing for such measly funds as are meted out. We have risen in the developed world to a state that, if things are decently managed, there is a surplus of everything. This was brought about by science; it should continue to be subsidized at a much higher level than that to which it has fallen, but not (only) because it is going to help us build a better TV set. Indeed, we should direct as many resources as can be marshaled fighting the longer term but lethal effects of global warming and industrial waste as a minimal custodial action for future generations.

VII. The idea of progress

What constitutes major progress, how is it recognized, what does it replace? Some progress is discontinuous, most has at least some precedent. This is not a history of Physics, so I will not go back too far, but if one were to trace a line from the 1800s, what points could one think of? The field broadens immensely thereafter: there are many directions and many strands of progress. By then, the full content of Newton's Mechanics and Gravity was fully digested and generalized by Laplace, Lagrange, Euler, Gauss and a galaxy of others. Now it was the turn of electricity and magnetism to go beyond the compass and the primitive electric batteries of Volta and company. The breakthrough was Øersted's 1820 discovery that two hitherto separate forces, electricity and magnetism, were in fact part of a wider complex. This first unification was completed with Maxwell's equations of the mid-1860s, while Faraday's earlier invention of the field was in some ways even more spectacularly applicable. Then came the evolution of the first and second laws of Thermodynamics and of statistical Physics in general, which focused on a completely different story. This was, in modern speak, emergent from mechanics. The electron was discovered in 1897. Then came Rutherford and Bohr, with the atom and the nucleus, the new quantitative notion of discrete fundamental entities in terms of which all others could be built. At the beginning of the 20th Century then, came the whole cascade of discovery of the very small: first the plum-pudding atom, then the concentrated nucleus, the notion of separate elementary particles, the proton and electron. The overshadowing event was the 1900 discovery of the Quantum by Max Planck, in connection with the totally different ultraviolet catastrophe in Classical Electrodynamics, from which one would never think that the whole rest of modern Physics sprang. In that context, it is amusing to see how Planck, the ultimate classical physicist, not only fought tooth and nail against anything too radical, but in so doing, found the most radical thing of all. He thought he was providing a purely stopgap "phenomenological" explanation, just a crutch until a better explanation could be discovered. He was also the one who said that new physical theories get accepted only as people die off.

I believe great physicists are, like Moses, fated never to reach the Promised Land, but only to lead us to it. I've seen that in so many scientists who made a big discovery, then just refused to go all the way it would take them, because it would lead beyond their notions of what was rational and feasible. Indeed, towards the end, my own mentor made his motto, "If you can't join 'em, beat 'em": if current developments do not look rational or useful to you, then you have to strike out on your own, however crazy that may seem.

Needless to say, the only matches to Planck's 1900 revolution were Einstein's of 1905 and 1915. In particular, consider 1905's three papers, each of which revolutionized Physics: Special Relativity (also, like Planck's Quantum, triggered by Electrodynamics), the photon as the light quantum (ditto)—observed by the photoelectric effect, and Brownian motion as the window into the molecular world. Special Relativity does have antecedents: these things are not called Lorentz transformations for nothing. However, Einstein was undoubtedly the one who made conceptually clear what Special Relativity is. All this while looking at other peoples' patent applications—a fairly impressive feat. The only thing that would trump that would be his discovery of General Relativity in 1915. That was the other summit of Physics. It must have taken courage after each of these revolutions for anyone else to dare to even enter the field. Nor must it be forgotten that Einstein was a leader in all aspects of the Quantum revolution well into the late 1920s, from his A and B coefficients to Quantum Statistics! Niels Bohr, who with Einstein was the greatest of them all, invented the hydrogen atom—quantizing atomic systems. His discovery culminated in 1925: QM out of Quantum theory.

In judging truly great breakthroughs versus less sensationally major advances in our understanding, it is important to separate the excitement of the moment or what appears to be big, from a longer-lasting effect, which takes an equivalent time to be fully understood. As we just saw, the 20th Century had more than its share of such moments. In 1905 these moments came thick and fast: they almost all seemed to be due to Einstein, standing on the shoulders of Planck, Lorentz and Poincaré. A recent semi-popular book, *Einstein and the Quantum: The Quest of the Valiant Swabian* by Yale's Doug Stone, showed Einstein's contributions to Quantum theory and

Quantum Mechanics over the years. The book details all the things that we didn't know or forgot that Einstein accomplished, cleared up and pushed forward, while his interests were focused explicitly on QM. I don't mean the later controversies about believing or not believing QM, but the explicit progress in technical understanding. By 1925, the revolution of Heisenberg and Born, Dirac, Schrödinger, Jordan, Pauli, and Klein, was one such milestone. Pauli's exclusion principle certainly was another, along with the notion of spin, culminating in Dirac's relativistic theory of the electron. That was one of the single most amazing feats of the human mind. Dirac was one of the Martians, who in a few short years changed the face of Physics, from the Dirac equation to his version of QM to Quantum Electrodynamics. He did this almost single-handedly. Such people gather legends; his stories are countless. I knew him more than I was entitled to, because he had shifted his interest to General Relativity about the same time that we at ADM got to work. Dirac didn't make it to the Promised Land: Quantum Field Theory, the very rock on which he had built Quantum Electrodynamics, was the most successful single human achievement ever with accuracies predicted, calculated and then verified experimentally to one part in twelve or more decimal places. Yet he no longer believed in it because of its associated infinities. In our last conversation, he asked me what was new and I replied that a new theory had been discovered to be finite. "Impossible, what is it?" "$N = 4$ super Yang–Mills" "I don't know any of those words." After I had gently explained that he sort of did know, his skepticism was unabated: he stated that if so, it must be a free field, and there we ended.

This exchange reminds us that in Physics there is no such outcome as a final theory. We understand things only up to a certain scale, beyond which we have no idea for the moment as to what might lie beyond. This whole scaling story, which began with the renormalization group of Gell-Mann, Low, Petermann and Stueckelberg, was brought to a profound and very satisfying level, in terms of our ignorance, by Ken Wilson, Michael Fisher and others. It states that it is really turtles, but very *different* turtles, all the way down!

VIII. Publish or...?

There are many byways in research. Even the great ones, like Einstein, also published smaller, less important works for a variety of occasions, such as solicitations from academies, historical schriften or because they just wanted to put something on record. There are those who believe that one should publish extremely little—only the very best of one's original ideas; they do not publish anything that could already be in the general consciousness or easily obtainable. When I was a graduate student, faculty members would gently say, "We don't publish everything we know." I, rather, believe in publishing "everything", but not more, than one knows. It does no harm: with online publications, it doesn't clutter journals or take up referee time or otherwise waste any energy but one's own. Perhaps it can serve, if only as a reminder to oneself, of a particular alley of research that could have been pursued further or was closed off by objections it raised. There's no prescriptive formula for what's right. It is also a matter of productivity coming, as Newton would say, in "fits of easy transmission and reflection." Some periods enjoy fertile circumstances; in others, life is difficult. For example, since the Standard Model was completed some 45 years ago, there really has not been any major progress. Instead, (observational) Cosmology has blossomed. There are different periods when different things move together, although there is a natural "anthropic" feeling that 40 years is a human research lifetime, so one would expect new ideas to come up as younger and newer creative activity comes into play. [In any case, I do not see much value in an anthropic argument. On the one hand, it hardly requires the whole standard model— unless some as yet unknown consistency condition (or say confinement) says we have buy the whole quark and lepton etc. package even if we need only the lowest set—and on the other, the existence of a sun-planet system just like ours can never be predicted from the big bang, only some vague likelihood estimate at best. The word reminds me of Fred Hoyle's dictum, "let's call a spade a horticultural implement".] There are no guarantees; we may have just been on a roll for a lucky few hundred years. Who knows when the next one will come?

Meanwhile students have to be trained and postdocs have to be helped towards interesting questions. There is never any lack of topics that one can dream up: it is just that they are not all of equal promise or fertility. Supervision of graduate students is a major task in the propagation of our faith, so to speak. It requires quite a bit of work on the part of the supervisor. Or you can suggest to the hapless candidate something that you cannot do yourself, or are finding heavy going, and see how it all turns out. Sometimes it turns out unexpectedly well, especially if this graduate student is called 't Hooft and is told to look at Yang–Mills renormalizability. At other times, it crushes the poor candidate who can't make heads or tails of what he could possibly be doing next. That is the process of apprenticeship and training and all the stages of reaching mastery in a profession, with absolutely no guarantees and no promises. That is something we have to bear in mind, both from the supervising and from the beginner's side: it can't avoid being difficult, but must not be impossible—a fine line to follow.

IX. Collaboration

Another interesting sociological subject is that of collaboration; it is different in Theoretical Physics, where large teams are not necessary, unlike experimental groups, which can comprise thousands of people. The image of the lone scientist sitting in his room or lab and spitting out wonderful theories is rarely correct, either. There are frequent meetings among scientists because they really need to keep up with each other and be encouraged by the activity around them. Collaborating and knowing, on a large scale, who is doing what, is important. For fun, I looked up how many different people I had written papers with. To my amazement, it's about one hundred, spread across many countries, mostly the US and Western Europe, but also Uruguay, Israel, Turkey, the Soviet Union in its day, India, Japan, both Chinas and Australia (two of my best postdocs—Jim McCarthy and Andrew Waldron). It does not mean that the work was done in those countries or indeed, in either of them. Several people that I have co-signed a paper with I have never met and am unlikely to do so.

It is a sacred law in our field that authors be listed strictly alphabetically. Under no circumstances, unlike in other fields, do senior authors come first just because they head a lab. It is usually the most obscure grad student, if he or she is lucky enough to have a name that starts with an 'A' or two (if Danish); too bad if you are called Zygmund.

X. Motivations

There are many peripheral questions regarding the behavior of scientists; one is their public attitude to big questions. By that I mean not the usual newspaper silliness about religious beliefs or lack thereof, but the presumably deep reasons we all went into the subject. It is mostly true that those of us working at the frontiers of basic Physics came into it with a desire to understand the universe we seem to be living in. Different people have different paths to the same end. After a while you become what you do, as in all phases of life, in this case, attempt either to solve problems within existing theories or go beyond and explore in various directions around an existing one. It is also striking that when speaking with colleagues at meetings or informal gatherings, the big questions are almost never broached. You do not approach an eminent colleague and ask, "What is the end-point of the universe?" or "What do you think we will be our grandest unification?" Instead you talk about concrete observations, experiments, theoretical results and non-results, but the talk is always technical and down to earth, within the limits of existing frameworks. What you thought "doing Physics" was in junior high school still is, but you realize that Physics is a lot about the small steps. The whole idea of Theory of Everything is really in the realm of journalism. Many of us no longer believe such a thing is meaningful. Ken Wilson has taught us that what we are doing at any given point is testing models that are good up to a certain degree of accuracy and energy level. We measure progress in how far we roll back the limits of the parameters that we think define the theory. This is a much healthier situation. You can always tell crackpots because their papers forget these little questions and claim that they have discovered something that proves everything. Superstring theory has held the floor since the mid-1980s, although there is considerably less physical, as against mathematical, activity in this subject. Similarly, Supersymmetry and Supergravity are still guiding lights for phenomenological suggestions at the accelerator level. In any case, apart from whether they have any physical applicability, they are far from the sort of revolutions that were Quantum Mechanics and General Relativity. It is still very much an issue whether we are reasonably formulating the question of what unites those two. Usually the right answer arises once you have framed the right questions; it is by no means clear that we know what these are yet.

I won't discuss extra-rational beliefs within the framework of basic science. It always surprises me that there are even a few scientists active in established religions. But there are and always have been. Some of them are by no means bad scientists, quite the opposite, but for most of us it is not simply part of the language. Steven Weinberg insists that the more we learn about the universe, the less sense it makes. I agree with that sentiment, but it is almost a truism. The question is really "What possible sense could it make?" Or rather, that question is simply not meaningful. Making sense is a human construct. In physics, we have learned repeatedly that you get in trouble using normal language, one that is not appropriate to the dimensions in question. That is why scientific language and equations cannot be replaced by everyday terms. Postulating a deity is simply a confession of ignorance, replaced by a new word.

We understand that we can get an awful lot of insight, let alone direct predictions that are correct, from the flawed models of Quantum theory and Electrodynamics, even if we don't yet understand how to get beyond them. The Standard Model still has many deficiencies, including the fact that it contains close to thirty arbitrary parameters that have to be set experimentally. But this is dwarfed by the fact that a few quarks and gluons determine the structure of strong interactions. Just recently there was even a no-parameters lattice gauge theory calculation yielding the correct neutron-proton mass difference, a subject that had been kicking around forever. These amazing triumphs are part of an otherwise complicated and not very well defined substrate. I won't even mention the larger scale, where there is no ambiguity whatsoever in Condensed Matter and other parts of Physics, which have never failed us in any respect.

On the other side, there is the cosmological scale. GR is a classical theory; indeed it is classical geometry that it talks about. It can be very fruitfully formulated as a field theory of its own, as well as an example of Riemannian geometry. It has done wonders, not just explaining and reproducing Newtonian Gravity, but given it a rational basis, such as the equivalence between inertial and gravitational mass of all matter as an infinitely "aced" test of the theory. In agreement with Special Relativity, it has brought out the fact that gravitational waves exist, that is excitations in this geometry or its field theoretical version. There is no doubt whatsoever about their correct existence and properties, as demonstrated by the secular accumulation of data from the binary pulsar and now with the accumulating LIGO triumphs in observing those gravitational waves. So GR is verified far more than by the usual "four tests" of popular articles.

Every theory has its domain of validity. General Relativity gets a bit tricky on the full cosmological level. On the one hand, it has allowed for the possibility of space-time being not merely asymptotically flat, but asymptotically de Sitter or anti-de Sitter. It does seem to be asymptotically de Sitter, despite theoreticians' preference for the anti-version. Aside from the deep question of compatibility with Quantum theory, there are still many unknown classical aspects: everything from cosmic acceleration in addition to expansion, as well as so-called dark matter. This is another word for observational deviations from the expected gravitational interactions at the galactic and cosmic level, which has proved all-too-fertile ground for crazy theories. I believe it can still be tamed within the Einstein framework, both for Occam's razor economy and from a number of promising results on that side, versus total absence of any observational evidence for it.

XI. Mathematics and Physics

There is no question that Mathematics is the language of Physics, nor is that a surprise. What is a surprise is that at every turn in physical theories there was available, or there became available, a mathematical tool or set of notions which miraculously seemed to fit the needs of that particularly evolving theory. The number of times it has happened is manifold. In the early days, Newton and others had to invent their own Mathematics, but then the line was not very pronounced between the two subjects. Calculus and geometry were the bases of Newtonian Mechanics. Differential equations were the language of both Classical and Quantum Physics. Analysis, functions of a complex variable etc, were immediately put to use in tools of our understanding of theory. It certainly seemed as though there was a miraculous fit. Sometimes, there is and sometimes there isn't. Sometimes we in Physics make up our own Mathematics or grope our way towards a certain set of ideas, which mathematicians then pick up on and do a much better job developing. There is certainly that correlation, and it was always the case, but perhaps a little more nowadays. Physicists have become more sophisticated and have acquired as part of their training, in some ways perhaps too much, mathematical knowledge and understanding. I'm always amazed at the concepts that seem to be altogether familiar to younger people in the field, yet not long ago were at the far edges even of mathematical progress.

That brings me to the particular notion known as the no-go theorems. Physics is not an axiomatic science, so there is no such thing as certainty or rigorous proof of anything, including the Sun's rising tomorrow. We are happy to have someone prove that some particular notion simply cannot be consistent with the existing canon. These no-go theorems are such a category, as the name implies. No-go theorems have a particular fascination for physicists because they immediately taunt us into attempting to circumvent them, which we usually succeed in doing. That is their beauty. Supersymmetry and Supergravity are such examples. They arose despite a particular no-go theorem which was practically invented to stop them. The nice thing about no-go theorems is they actually force us to think harder about what we assume to be physically necessary and correct. Another case is the famous CPT theorem and its possible violations. The CPT theorem itself is an enormous achievement. It arose quite early in the history of Quantum Field Theory, although the proof took some time and was elaborated at different levels of rigor. It states that no local Special Relativistic QFT violates conservation of the product of charge conjugation, parity transformation and time reversal (CPT). That led to the (unjustified) belief that not only did the product of these discrete transformations have an inviolate status, but so did each of its three components. That turned out to be spectacularly wrong. Parity violation was predicted and observed and led to the 1957 Nobel Prize for Lee and Yang who showed that the "paradoxes" of weak decay were not paradoxes, but simply "forgotten" Parity violation. Conversely that meant there was also CT violation because a product had to stay invariant. Later, there was the famous discovery, without any theoretical suggestions or input, of loss of T-time reversal invariance. This remains one of the great mysteries. Different interactions are responsible for different violations. Parity just means mirror symmetry. Time inversion symmetry at the layman's level certainly seems very unlikely to hold because we don't believe that the laws of Physics are symmetrical in time, but they certainly are in the sense that running any movie backwards results in a movie that could have happened, however unlikely. Charge conjugation is a bit more esoteric: interchanging electrons and positrons. The basic CPT theorem has held, although it is not as if we cannot construct theories that violate it. But we know exactly what the price is and so far that price has not proven necessary or desirable.

At the end of the day, although we would like to say that elegant theories are bound to be right, that has not always proven to be the case. Indeed, even Supersymmetry, Supergravity and Superstrings, beautiful and almost inevitable as they may seem, may not be the next rung in our future progress. The case is not yet closed nor clear, but on the other hand there is no doubt that, whatever the next step will be, we will decide that it is elegant. If a theory works, we eventually decide to confer elegance or discover why we should have known it was elegant in the first place. Areas that are nowadays considered simple like vector spaces with complete sets of functions or indeed the notion of Hilbert space were not always so. There is always the creative tension between knowing too much math, the part that is not going to be useful, and knowing too little math and not understanding what one has stumbled upon—as Heisenberg did in 1925, not knowing matrices.

XII. Humor

Even physicists have a lighter side and generate aphorisms and jokes, although they may not resonate with non-scientists. We all know the many quotes from Einstein; more are ascribed to him than he ever said. He claimed he was a straightforward man of plain stock, although his remarks hardly reflect broad peasant humor: *"Raffiniert ist der Herr Gott, aber boshaft ist er nicht"*—"Subtle is the Lord, malicious he is not" can hardly qualify.

Niels Bohr had his own stories, well-known in certain circles, but perhaps less so than Einstein's. One concerns superstition. A visitor was aghast that in Bohr's summer cottage there was an upside-down horse-shoe over the door. The visitor said, "Professor Bohr, surely you don't believe in that?" Bohr replied indignantly, "Of course, I don't. But they say it works even if you don't believe in it." That particular twist of humor has descendants. For example, Quin Luttinger said, "Happiness isn't everything. Look at Roy Glauber, who is perfectly happy about being unhappy."

Wolfgang Pauli never really quite grew up. He was a *wunderkind* of the most egregious kind. At age 18 he wrote the first textbook on General Relativity, soon after its original publication! He had attended a Viennese operetta, something like *Die Fledermaus*, where a maid looks in on a ball for the nobility and is shown the heir apparent. The maid exclaims "so young and already a Prince!" Pauli loved versions of that quote: about some of the junior people: "So young and already unknown," or one that I witnessed about Gunnar Kallen, "So young and already a Privy Counselor." There were many jokes about Pauli and the Pauli effect—that whenever he walked into a lab things stopped working. One of them was that when he was going to visit a particular lab, they had put a bucket of water above the slightly open door in order to cause a (rigged) Pauli effect. So he opened the door, and sure enough the bucket did not fall.

Unfortunately, the use of humor in scientific papers is all-too-dampened by referees and cautious journals. When Eddington had gone off the rails and come up with his crazy Theory of Everything in a book called "Fundamental Theory", he would invoke incredibly large and small numbers, then correct them as needed later. A paper by Gamow and friends that slipped through the censors proposed that there were so-and-so many particles in the universe. This was done as a set-up to print a later correction stating, "Unfortunately in one of the key formulas in our previous paper, the sign of the exponent, instead of being 10 to the plus one hundred should have been ten to the minus one hundred. This, however, doesn't change any of our conclusions". Of Dirac, I cannot even begin! We are much duller these days.

XIII. Refereeing and other chores

One subject that is a little offbeat, but occupies significant time for many of us, is refereeing for journals and grants. The situation has changed considerably from the old days when there were few journals and, within these few, a hierarchy; now there is an infinite number of journals and it is guaranteed that any paper, however self-contradictory and wrong, will be published somewhere because all the hungry publishing companies are set up simply to make a few rupees or dollars, forcing libraries to subscribe to them. These journals need to fill their pages, so even if they go through the motions of refereeing, sooner or later there will be a referee who will okay a paper. Most of us, especially older ones who grew up with the system, knew that it was a tax to be paid in order that one's own papers also got decent refereeing. As Yogi Berra said, "Always go to other people's funerals, or they won't come to yours." On the whole the system has worked. There are certain tacit rules. I remember when I was fairly young and still amazed that I would get *Physical Review* and similarly respected publications to ask for my opinion, having to referee a paper from a distinguished Nobel-Prize winning author, which was partly wrong. My attitude then, as it would be now, is that one should point out what the shortcomings are, what the wrong steps might be, but at the end of the day leave it to the author to decide. Einstein actually submitted a wrong paper, in his early days in the US, to the Physical Review. He got a very detailed and useful report from the referee as to what the errors were. He replied in a huff, "I didn't submit this paper to be refereed, but to be published." That was more or less the German system in those days for certified geniuses. He never published in the Physical Review after that. I've been on both sides of that one, too. I have received a report with the referee saying "If he wants to publish this, fine, but let him at least think twice before doing so." It is this sort of interaction that keeps our subject in reasonable shape. No system if foolproof. No refereeing is objective, pure, or infallible. A good referee will notice something that the author should have done, and a bad one might fail to notice what is truly novel and right. To this day, distinguished, as well as run-of-the-mill, theoreticians, get calls for refereeing. Some people feel themselves too high to bother, while others, after many decades of refereeing, have certainly earned the right to withdraw from the system. There is a feeling of curiosity and to some extent the feeling "Let me see if I can reasonably quickly figure out what is going on." Mathematicians are different. Sometimes their papers take years to be published, because the referees take such painstaking work to check every step of a proof. Theirs is a very different game. There are also times in which a referee report will discourage the author of a possibly revolutionary, or at least novel, road from taking it, but that is less likely.

In the early days of Quantum Field Theory, Jordan invented Fermi–Dirac statistics, submitting his manuscript to Max Born, then the editor of *Zeitschrift für Physik*. Born was just leaving for the United States and put the manuscript in the bottom of his suitcase; by the time he came back, Fermi and Dirac had independently come to understand the notion. So poor Jordan was out of luck. That has happened in another classic, but little-known example: Oskar Klein independently invented the before Schrödinger did, but then contracted jaundice; it nearly killed him, and kept him six months in the hospital. By then Schrödinger had put out his paper. Conversely, Schrödinger apparently found the Klein–Gordon equation first. There are also cases in which the person publishes first, but does not even belatedly get credit. Gunnar Nordström, a Finnish physicist, discovered the notion of higher dimensions in 1914. He, too, was trying to unify Electromagnetism and Gravitation, but he did it the pre-General Relativity way, as he only knew that gravity had to be Lorentz-invariant, the simplest choice then being a scalar field. He invented five-dimensional space. Four of the metric components were the electromagnetic vector potentials; the fifth, scalar gravitation. We know from the fact that light bends that gravity must be a tensor rather than a scalar; nor can it be a vector, because vectors lead to repulsion between like masses. That was what led Klein to unify, again in 5D. This time 4D was tensor gravity and the fifth the electromagnetic field. He too got a scalar field at the corner. Nordström and Klein made a magnificent generalization. To this day, the notion of higher dimensions continues to haunt us; every physicist finds it a truly intriguing idea. We know that Superstring theory is consistent in ten dimensions. Supergravity goes up to eleven dimensions. There are many places where higher dimensionality kicks in, even including attempts at infinitely many, which however have not gotten far.

I started to talk about refereeing, but ended up in (unwarranted?) generalizations. They all belong to the same basket of groping our way towards the future. The French poet Guillaume Apollinaire wrote, "For us who are struggling in the trenches, have pity on our errors, pity on our sins." He didn't write this about QM, but it is certainly very apt for those who are on its front lines. One should understand that there is nothing wrong with exploring a little too far and a little too crazily, as long as one distances oneself from the more egregious paths. It is all-too-possible to get so captivated with an idea that is wrong and to give up everything else. Yet without craziness, we will not have the progress we need. So I am in favor of craziness and in favor of refereeing to keep track of that craziness.

XIV. Vocations

There are many roads to becoming a scientist, let alone which branch of science and why. There are probably as many reasons as there are practitioners. This may be of some interest to students or others groping towards such a lifetime choice. In my case, curiosity about the universe came around age 12 or 13, just about the time I realized that a superhuman being having done it all was not an answer but begged the question. It was only when I discovered books about popular science that I realized that people had really thought long and hard about this. In high school, I was still not aware that systematic understanding existed. Even as I came to college, I assumed that Physics was a static, completely accomplished discipline in which people understood once and for all the questions that are raised by wonder and looking up at the sky, let alone apples falling. The ideas that it was possible and desirable to devote one's life to the pursuit and that there was anything left to do came very slowly to me. That there remained mysteries was not something you learned as an undergraduate, since you were learning about Maxwell's equations or some elementary Quantum Mechanics, which were known to be correct. Even the idea that there was something called graduate school was a strange notion. I could understand that there was more to explain, but it was still not very clear to me what people did in graduate school. It was still not clear once there, especially if you have brilliant professors like Schwinger who present a subject as all wrapped up, like Quantum Mechanics, whose details are completely understood and known at that level, apart from philosophical controversies. There seemed to be nothing left to do. Gradually I began to perceive that there are, and will always be, wide-open subjects. Not only that, but something to which one can contribute original thoughts. It is a very arduous process, except I suppose for the Martians, who innately understand everything about what is and what one should try to accomplish. For the rest of us, the work is full of dead ends and walls to bump into in an attempt to accomplish something.

The first original idea, and the first paper I ever published, was derivative; with Paul Martin, we extended Schwinger's two-body (AKA Bethe–Salpeter) equation. It was a little excursion into Quantum Field Theory. This tiny advance in a different direction did not lead to much wisdom, but then in the long run, neither did the two-body equation. At some point I began to realize that I was on my own. I was at a border where there was no textbook and no paper that would tell me what the answer was or what to do or how to do it, which is equally important. It is extremely difficult to learn when a problem is too difficult and relatively speaking, insoluble, at the present stage of the art. Indeed, it is very important to jump off those questions, otherwise they will swallow all one's time. We all occasionally solve problems that are too easy, simply for the fun of doing them. We all exposit at some point: after all, we teach courses. We also give review lectures at meetings and communicate the state of the art, which necessarily involves a certain amount of summarizing and putting together what is known, but in a slightly different way. We do all that, but the primary push is to come up with original ideas that are doable but not too doable, and yet provide satisfaction. If you never succeed, of course, you risk disaffection.

The attrition rate is quite high. After a PhD, the number of people who keep publishing, stay in the field and do meaningful work, is a fraction of those that earned the degree. Eventually, everything limits to zero. You are going to stop doing interesting work. The questions are: How long? How well? It depends partly on one's tolerance level, one's feeling of ability and simply the fun of doing things. By definition, a vocation means that there is nothing else that you would want to do instead. It means that you really breathe, think and sleep your subject almost exclusively. In your early years, which can be some of your most productive, it is unlikely that you will go far if you fritter away energy and time on other pursuits, be they finances or travel. Nobody works non-stop, but automatically your thoughts will return in the direction of what you are doing. It is not just an academic year pursuit: summers are when faculty can get their best work done.

Suppose you have finally finished a problem; there is the pleasure of having done, mitigated by the exhaustion you feel from having gone through it. You often feel like walking away and moving on to something else, rather than publishing some version of your results—which is part of your duty to your community, so they won't be needlessly duplicated and will prove of use to other people. There are people who don't bother publishing what they have done; it is too much of a distraction. It is also true of some very great scientists. What they have published is only a fraction of their accomplishments. I don't mean Newton's simile of walking along the seashore and picking up cute little pebbles instead of looking at the ocean of truth. In the old days, there were so few scientists that they would of necessity write to each other just to have some reaction and some idea of whether what they were doing made any sense. That, indeed, required a certain amount of concentration to describe the results. Now that it is almost unnecessary, we still do it.

There is a wide difference in our presentations. First of all, the prose: English has become the default language in which everyone writes. People for whom English is distant write almost illegible papers. There are also plenty of native English and American speakers who write as if they don't know the language. Others write in such a polished way that you are carried along by the prose. It is an evolving, rigorous, difficult and frustrating process. Despite the difficulties, the satisfactions outweigh the disappointments. That is true at all levels. If you are Einstein, there is no need to say anything about the process. For us mortals, of whom there are many degrees, you become what you do—in any discipline. The pleasures are there and genuine even if the results are considerably more modest than they would be from someone of higher talent. There is a natural match. People whose ambitions are bigger than their abilities, which includes almost everyone, will find life frustrating. We all have those ambitions; how we react to them varies. There is no optimal prescription. You feel your way, you get various positive feedbacks and, like any lab animal, you tend to do things that you found pleasurable in the first place. There is nothing more so than solving a problem, even if it wasn't the world's greatest. Even if it wasn't the problem that you started thinking of solving, but to have added your own little coral to the atoll reef is a source of great pleasure. There are added pleasures, like the endorsement of your peers, though you know deep down the value of what you have done. Still, it is nice to have it endorsed. As with any vocation, anyone who makes it into that particular subset of investigators who really profit from and contribute to the field, there are very few joys that can match creation. I mean that in a wider sense; it carries over to the arts as well. Scientists, especially physical scientists, are often accused of not paying attention to the humanities or to other developments of the mind. That is nonsense. Most of the good scientists I have known were cultured, able to appreciate other intellectual endeavors. Physics is indeed a jealous mistress, but not to the extent of forbidding everything else.

XV. Lastly, the basic Physics questions

This is perhaps the hardest, and maybe best skipped, part, but having alluded to it so often, I'll have a try. First, we must remember that the language describing our knowledge of the very big and small is a different tongue from English, and it has roots in our primitive six foot (certainly less than say 1000 miles and more than a millimeter) and one second to one hundred years, dimensions (not to mention less than say Mach 3 speeds and gram to ton masses). There never was any reason to think that this language would extrapolate to numbers different by factors of trillions! Yet, as humans we have to express our results in our lingo. First, *"Why is there something rather than nothing?"* sounds perfectly reasonable—but is it? I will preface that with the old Yiddish story of a Council of Learned Rabbis (or whatever sect you prefer), who conclude their discussions by saying "So, we agree that not having been born is the best fate, but who can be so lucky—maybe 2%?" Those 2% must be asking *"Why is there nothing instead of something?"* It is not a logical question, because there can be no logical answer—the notion of a Creator just pushes it back one step—while saying that's what the laws of nature require is again just a rephrasing. So one must accept the existence of our observed universe as an axiom, just as Mathematics requires an axiomatic base. The more modest "Why are these laws the way they are?" seems a bit more promising, but still requires care. Something called the Anthropic Principle, whose gist is that whatever they are, they must allow us to exist, is in that sense a truism. Still, within our broad set of laws, even a slight change of value of some numerical quantities, or of some obscure energy levels of the iron nucleus, would easily forbid us. Furthermore, since our laws can only be known to hold up to some energy scale means we should not get too metaphysical about it all on the basis of our current ideas. But of course we do! As we now understand nature's laws, they consist of certain equations which admit of many solutions, depending on "boundary" or "initial value" conditions, a simple enough concept already at the Newtonian level. His laws of motion only fix a ball's or planet's trajectory if we specify its position and velocity at some moment. More ambitious attempts at "no boundary", or boundary conditions included in the laws, have also been recently proposed.

A perfect illustration is the universe itself as the ultimate system. Our current observations seem to point to Big Bang (so-named as a term of derision by its opponents, incidentally) Cosmology as a unique starting point. It is *not*, I repeat, a little dot in some background space that suddenly zooms out—there was no "there" there, no before, no awakened sleeping space! It permits orgies of speculation about the accompanying trillions upon trillions of multiverses that bubble off in all directions, like the arms of those Indian Gods, each with its very own laws of multi-nature, for all we know. We are then (some of) the lucky ones, living in an anthropically favorable world—by definition. Indeed, one Big Bang version has these expansions ultimately evolving into the movie running backwards to a collapse, whose new Big Bangs could be entirely different. One can see those medieval angels dancing gleefully on a—very shaky—pinhead! A Big Bang is by no means required to support inflation, its original *raison d'être*. Very recently, for example, much less violent universes have been constructed, ones that expand and contract, perhaps indefinitely (so no initial condition needed), but do so in a gentler manner, rather than down to the Big Bang's point-like throat, so no multiverse or recycled laws or other Sci-Fi scenario; a duller, but perhaps more satisfactory, ride! This is by no means the end of the story, but it shows concretely, in terms of current observation, that premature metaPhysics need not apply. The very use of ordinary language, evolved in pre-scientific times, often leads to pseudo-problems. A current example is the emergence versus reductionism controversy. The latter states that everything can ultimately be understood in terms of the Standard Model, whereas the former states that novel, irreducible, properties emerge at higher levels, for example, in Condensed Matter Physics where very many elementary particles participate. But this is no contradiction, just one of language use: in principle, all can be reduced, in the sense of being calculable, but it is impossible in practice. Again, a category confusion, rather than a fundamental dichotomy. Each level of understanding or description has its own natural categories.

Next, is there, as Einstein especially hoped, some physically unique set of laws, even if we are far from having found it? This might mean that whatever the improved laws at higher energies, their limit at our lower resolutions is unique. These questions seem to be, unlike some of the non-questions above, perfectly reasonable ones, perhaps within some added anthropic proviso as a worst case. We just don't know—yet. For example, Superstring theory in ten dimensions might be the only finite consistent Quantum theoretical model, but so far it has not yielded any limit that looks like our four-dimensional world, with any of the particles we know to be present in the Standard Model. I don't think questions that deep will be resolved any time soon, at least until the arbitrariness of the many parameters in that model itself is resolved (but before you bet, remember Bohr's 1924 despairing words that it might take a century before a QM revolution!). In any case, the revelation that the laws of Physics gradually improve in scope with our vision, that is, our higher energy insights, teaches both modesty and ambition. It's the modern version of Plato's cave: the shadows on our walls become sharper as our screens become more fine-grained!

This discussion is complex enough to deserve another phrasing, identical in content, that may be a bit more enlightening. The seemingly grammatical question "Why is there something?" is no more meaningful than "Will you rain?". There is no possible finite regress of answers to provide the "why", rather, it will just be replaced with another "why?" A related quasi-problem is "Why is the world intelligible?" Even apart from defining the acceptable level of "intelligibility", or indeed begging the question with the word "why"; an anthropic answer is that for us to exist, a certain stability of the perceived universe is required, and any stability can be approximately captured in an empirical set of "laws" that themselves change as we learn more, without being in any way final. At our present stage, this set consists of GR, QM and the Standard Model—within a certain observational band—with many open questions still bedeviling this level of intelligibility. So, at best, we can render somewhat understandable a bounded, still-hazy, set of observations, but must accept much deeper ignorance of each of its components—the Standard Model's many free parameters and incomplete field theories being salient examples, let alone cold dark energy's dynamics—indeed, our less charitable successors might not credit us with having found intelligibility as much as declared it by hand-waving! Even if we have a credible toy model of some universe, it may vastly differ conceptually from that of the true(r) universe(s); indeed, we are not even assured of being smart enough to understand the latter, hence the questionable "why" of intelligibility! Clearly, a lot more coral deposit is required before the atoll, if it even exists, is complete: we are unlikely to paint ourselves out of our vocation any time soon!

ACKNOWLEDGMENTS

Why There Is Something Instead of Nothing

Even as modest a volume as this owes its existence to many people, in addition to its dedicatees. Historically, I was prodded into it by several colleagues, especially Thibault Damour and Tony Zee, who felt that my witnessing the last of the creators of modern Physics was grounds enough. My wife and children's suggestions proved invaluable; Abigail's devotion and wise editing should be thanked by all readers—she deserves all the credit, none of the blame. Katie Clark had the unenviable task of turning my initial oral history into a first text. Benjamin Schwarz's insightful questions and comments sharpened my vision. To the World Scientific Publishing staff, in particular my editor there, Annabeth Zhang Lu, and its founder Dr. K. K. Phua, my sincere gratitude. Finally, the Lounsbery Foundation and its officers' ongoing support and trust that something might yet emerge, was priceless.

IMAGES

Polish passport photo c. 1934 with my mother Miriam (nee Melamed).

My Uncle Emile, Miriam's brother.

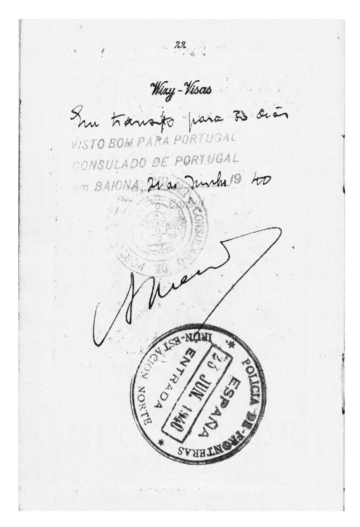

Portuguse transit visa signed by Consul Sousa Mendes and Spanish entry stamp, 1940.

COMPANHIA COLONIAL DE NAVEGAÇÃO

S. A. R. L.

SEDE NO LOBITO

Talão que o passageiro deve conservar

Bilhete de passagem de _____ classe

Paquete: GUINÉ

Viagem N.º

NOMES	Edade

FARE(S) COLLECTED Esc.

Adultos	Meios	Quartos	Grátis

De _____ para _____

a sahir em ___ de _____ de 194_ às _____ horas

(salvo caso de força maior).

Camarote N.º _____

Beliche N.º _____

Sofá _____ de 194_

Pela COMPANHIA COLONIAL DE NAVEGAÇÃO

Os serviços clínicos, medicamentos e dietas serão pagos pelo passageiro em conformidade com a tabela de bordo.
Os Srs. passageiros devem estar a bordo duas horas antes da indicada para a saída.
Os bilhetes são intransmissíveis e ficam sujeitos ás condições exaradas no verso e mais regulamentos de bordo.

Ticket for passage aboard the Guine, Lisbon to New York, 1941.

Graduate student, c. 1949 (?)

With my mother, Princeton 1953.

From my postdoc years.

In the Morgan with Elsbeth, Denmark 1955.

Our 1956 Copenhagen wedding day, with my father and the Kleins.

We reach the Arctic Circle, 1958.

Camping near Nordkapp with the famous Deux Chevaux.

Talking Physics with my father-in-law, 1960s.

On the banks of the Charles with Kleins, c. 1965.

Arnowitt, Deser and Misner (ADM):

From a local Danish paper, 1960.

Oskar Klein Centenary, Stockholm 1994.

My sister-in-law, Professor Birgit Arrhenius, nee Klein, mid-bottom row.

Signing in to the National Academy of Sciences after election, 1994.

Caltech, 2010s.

A recent snapshot.

With Abigail, my editor-to-be, c. 1967.

INDEX

CPSIA information can be obtained
at www.ICGtesting.com
Printed in the USA
BVHW012145220921
617383BV00009B/193